A Language of Things

Studies in Religion and Culture

John D. Barbour and Gary L. Ebersole, Editors

A Language of Things

Emanuel Swedenborg and the American Environmental Imagination

DEVIN P. ZUBER

University of Virginia Press
Charlottesville and London

University of Virginia Press

Printed in the United States of America on acid-free paper

First published 2019

9 8 7 6 5 4 3 2

Library of Congress Cataloging-in-Publication Data

Names: Zuber, Devin P., author.
Title: A language of things : Emanuel Swedenborg and the American environmental imagination / Devin P. Zuber.
Description: Charlottesville : University of Virginia Press, 2019. | Includes bibliographical references and index.
Identifiers: LCCN 2019015592 (print) | LCCN 2019021594 (ebook) | ISBN 9780813943527 (ebook) | ISBN 9780813943503 (cloth : alk. paper) | ISBN 9780813943510 (pbk. : alk. paper)
Subjects: LCSH: Environmentalism—United States—Religious aspects—History—19th century. | Swedenborg, Emanuel, 1688–1772—Influence.
Classification: LCC GE197 (ebook) | LCC GE197 .Z83 2019 (print) | DDC 261.8/8—dc23
LC record available at https://lccn.loc.gov/2019015592

Cover art: Detail of *Reverend Joseph Worcester's House, Piedmont,* William Keith, ca. 1883. (Private collection; image reproduced courtesy of Alfred Harrison and the North Point Gallery, Berkeley, California)

For Suzanne, Sophia, and Catherine

Swedenborg, of all men in the recent ages, stands eminently for the translator of nature into thought.

—Ralph Waldo Emerson

Contents

Illustrations

Acknowledgments

This project has been many years in the making, and I am deeply indebted to the intellectual generosity of so many who gave it their time and attention. The original seeds of the book lie in my 2010 doctoral dissertation, written under the direction of Joan Richardson at the Graduate Center of the City University of New York; Joan has remained an indispensable critical interlocutor (and friend) over the subsequent years: without her formative seminars on American Aesthetics, I would not have first tumbled down the rabbit-hole of attempting to trace Swedenborg's diffuse presence in nineteenth-century thought. That dissertation was awarded the CUNY Graduate Center's Alumni and Faculty Prize for Most Distinguished Dissertation, and I am very grateful to the CUNY English Department for the publication subvention that has helped bring this book (finally) to fruition with the University of Virginia Press.

A Language of Things ended up tacking a different direction than the transatlantic one originally plotted out by my dissertation, however, as the subsequent project came to gestate over the course of several rich fellowships and scholarly residencies that provided a different set of coordinates (questions, problems, provocations). The brunt of the John Muir material was begun while I was the 2010 European Fellow in American Studies at the Eccles Center at the British Library, London; many thanks to Philip Davies and the staff at the Eccles Center for facilitating that most fruitful stay before I moved to Berkeley. As a 2014 fellow with the Gardarev Center, the first draft of the Jewett chapter came together while in residence with Gardarev at the Mesa Writer's Refuge in beautiful Point Reyes, California—Toni Lester, Gardarev director, has always helped keep the subjects in this book "real," and grounded in the sociopolitical where it matters. While on sabbatical in 2015–16, three different institutional homes gave the book its final, critical shape: the University of Stockholm's Institute for Culture and Aesthetics, where I was a visiting scholar for seven months (particular thanks there to Inga Sanner and Staffan Bergwik); the Ingmar Bergman estate on Fårö island, which afforded weeks of uninterrupted writing, punctuated by long walks on the stony Baltic shore; and the Rachel Carson Center for Environment and Society at the Ludwig-Maximilian University of Munich (so many there to thank, but especially Christof Mauch for making my summer residency as a scholar feasible). Various parts of this book benefited further from invited talks at the following institutions and organizations: Urbana University (Ohio), Bryn Athyn College (Jane Williams-Hogan and Aram Yardumian), the University of Lund, Sweden (David Dunér), the London Swedenborg Society (Stephen McNeilly and James Wilson), the Skandinaviska Swedenborgsällskapet in Stockholm (Göran Applegren and Gustav Fernander), UC Berkeley's Botanical Gardens (Erin Wheeler), the Royal Swedish Academy of Sciences (Karl Grandin), the American Association of Art Appraisers annual meeting in San Francisco (Holly Mitchem), the Massachusetts New Church Union (Herb Ziegler at the Cambridge Chapel, and Susannah Curry at the Bridgewater New Jerusalem Church), the Fryeburg New Church Assembly, the Pacific Coast Theological Society (Ted Peters), the Pacific School of Religion's Summer Lecture Series, and the University of Oslo, as part of the Norwegian-funded REDO research network (Jone Salomonsen and Sarah Pike, among others).

Closer to home, my colleagues at the Graduate Theological Union and the University of California, Berkeley, have been essential in shaping the questions around "why Swedenborg mattered" for ecocritical appraisals of nineteenth-century American literature—my fellow professors at the GTU's Center for Swedenborgian Studies (CSS), Jim Lawrence and Rebecca Esterson, provided indispensable input and advice, and were quick to set me straight when it came to Swedenborg in the maze of his eighteenth-century contexts (any remaining errors regarding Swedenborg's biography here are wholly mine). Alan Thomsen gave keen feedback on parts of the book that dealt with Arts and Crafts history (and thanks to Francesca McCrossan, former secretary extraordinaire at CSS, for having given every measure of support in the background). I am deeply appreciative of the entire executive board of the Center for Swedenborgian Studies for not losing faith as this project stretched out over the years, certainly longer than they ever anticipated, and for nevertheless approving my extended sabbatical stay in Sweden and Germany. Other GTU colleagues, past and present, that I wish to thank for their perspicacity on different sections of the book include Rossitza Schroeder, Kathryn Barush, Naomi Seidman, Judith Berling, Rita Sherma, Christopher Ocker, Andrea Bieler, Chris Renz, Inese Radzins, Randi Walker, Christina Hutchins, Marion Grau, and Frank Burch Brown (who is last here but by no means least, for it was Frank, during his year-long tenure in Berkeley, who propitiously suggested that my project might find a home with the University of Virginia Press). The GTU's Center for the Arts and Religion (CARe) has been a consistent second home since I came to Berkeley, and I am so grateful for Elizabeth Peña's constant friendship and the awards committee at CARe, which kindly granted an additional subvention that allowed for the reproduction of some the beautiful paintings that appear here.

Additionally, a draft of the second chapter benefitted from being workshopped and critiqued in the Public Theology Inquiry Group at UC Berkeley's Center for the Study of Religion (BCSR) that I am part of, and I remain indebted to Jonathan Sheehan for that ongoing valuable space for intellectual exchange between GTU and UC Berkeley faculty—special thanks there to Charles Hirshkind, David Marno, Susanna Elm, and Niklaus Largier. Other individuals and colleagues who helped hone earlier versions of chapters include Leti Volpp, Richard Perry, and others who participated in an invigorating July colloquium at the Rachel Carson Center at the LMU in Munich; Michael Quick (who set me straight on Inness's *Home of the Heron*), Karen Victoria Lykke Syse, Johannes Voelz, Magda Majiewska, Karen Weiser, Suzanne Zuber, Amanda Davidson, and—especially—Anders Hallengren. Michael Jonik provided a last-minute rich exploration of Peter Lamborn Wilson's use of Johnny Appleseed and Swedenborg; Tyler Green also deserves a big thanks for a thorough, critical reading of the Muir and Keith materials quite late in the game, and for continuing to expand my understanding of Emerson's amplifications in nineteenth-century American art. I'm also indebted to the two anonymous reviewers of the manuscript at UVA Press who offered extremely valuable feedback and suggestions for improvement; remaining errors are wholly my own.

The GTU in Berkeley—"where religions meet the world," as our institutional tagline goes—is an extraordinarily diverse (and often challenging) environment in which to teach. I have learned from so many of the students who have persevered with me through Emerson, Muir, Jewett, and the other characters in this volume (including Swedenborg); these include Colette Walker, Austin Salzwedel, Meredith Massar Munson, Yohana Junker, Tyler Gardner, Jeremy Sorgen, Matthew Hartman, Luke Bruggeman, Cassie Lipowitz, Sarah McCune, Noah Lyons, Joy Barnitz, Cory Bradford-Watts, Jeffrey Adams, Colin Amato, and Claire Chuck Bohman. Further gratitude is due to those research assistants who fetched any number of books from the depths of the Doe Library, or (worse), were asked to transcribe the cribbed nineteenth-century handwriting of James John Garth Wilkinson in his acrostic-like letters to Emerson and Henry James Sr. (thanks here to Rob Peach, Colin Amato, Thomas Müller, and Tyler Gardner, again). The University of Osnabrück invited me to return as a visiting professor for three consecutive summers, where I was fortunate to teach intensive seminars on various materials from the book to a fantastic group of graduate students, including Lara Kötter, Chris Alcacer Labrador, and Jasmin De Nys. I am indebted, always, to Peter Schneck for making those rich stays with their rewarding conversations possible (which tended to linger late into long summer twilights, often at a restaurant on the beautiful square in front of the Marienkirche), and also to the others who have been part of the "team" in American studies there in different constellations (Sabine Meyer, Sara-Duana Meyer, Jens Bonk, Nadja Hekal).

If the seeds of the book were a dissertation, the roots that took hold are grounded in a number of archives and special collections connected to Swedenborg or nineteenth-century Americana. This book would simply not be without the able assistance of these places, librarians, and archivists: the special collection and archive at the Center for Swedenborgian Studies at the GTU (Michael Yockey), the Royal Swedish Academy of Sciences, Stockholm (Maria Asp, Jonas Häggblom, and Anne Miche de Malleray), the London Swedenborg Society (Alex Murray), the Bibliotek Swedenborg at the Swedenborgs Minneskyrka, Stockholm (Eva Björkström), the library and special collections of the William Keith collections at St. Mary's College Museum of Art, California (John Schneider), the archives at the San Francisco Swedenborgian Church (the Rev. Junchol Lee), the Bancroft Library at UC Berkeley, the Harvard University Archives, and the University of the Pacific's Holt-Atherton special collections (Mike Wurtz and Nicole Grady).

The book is dedicated to my best friend and fellow-adventurer on our swiftly tilting (and warming) planet: Suzanne Schwarz Zuber, who has been behind every chapter (even driving our car back from the Grand Canyon so I could finish drafting this book's afterword in the back seat). She has brought our two bright girls, Sophia and Catherine, to love this world with so much wonder—may they follow some of those roads not taken that this book tries to recover, and work toward a brighter future beyond the Anthropocene.

A Language of Things

Introduction
The Language of Things

> The translation of the language of things into that of man is not only a translation of the mute into the sonic; it is also the translation of the nameless into name.
>
> —Walter Benjamin

In 1940, disturbed by the growing cracks apparent in the façade of Western civilization, the German critic Walter Benjamin penned a series of sobering "Theses on the Philosophy of History." The twenty numbered paragraphs, dense and aphoristic, are driven by a messianic anxiety over the darkness then encroaching on Europe (it was to be Benjamin's last major piece of writing, before his failed attempt to flee Vichy France). In perhaps the most well known part of the essay, Benjamin turns to a small, beautiful painting of an angel by the artist Paul Klee (which Benjamin had acquired earlier, in 1921), in order to offer an ekphrastic description of the *Angelus Novus*—a "new angel," in Klee's titling—as a complex visualization of modernity's current predicament. "This is how one pictures the angel of history," Benjamin wrote,

> where we perceive a chain of events, he sees one single catastrophe which keeps piling wreckage upon wreckage and hurls it in front of his feet. The angel would like to stay, awaken the dead, and make whole what had been smashed. But a storm is blowing from Paradise; it has got caught in his wings with such violence that the angel can no longer close them. The storm irresistibly propels him into

the future to which his back is turned, while the pile of debris before him grows skyward. This storm is what we call progress.[1]

Benjamin might not have anticipated how his words would so deeply resonate with our contemporary environmental crisis, and the specific sorts of ways that the angel of history has accrued new ecological overtones, as Bruno Latour cogently observes.[2] From the piles of plastic debris amassing in the ocean to the rise of superstorms, spun off by increasingly destabilized weather systems, we find ourselves in a position akin to Klee's wide-eyed figure, entangled in an ecological violence that our technocratic society seems unable to unravel. The flood of doomsday news regarding climate change continues to grow, a litany of forecasts and scientific reports seemingly without (yet) any adequate redress by international legal frameworks, as the weak, nonbinding agreements that came out of the Paris climate talks in 2015 underscored. At that last, most recent gathering of world leaders, as carbon cap-and-trade credits were being haggled over, and one heard arguments about the need to not impede or hinder economic progress, the list of countries already facing climate refugees due to the rising levels of the oceans (including the United States) increased.[3] Shortly after the meeting, yet another international scientific study was released in *Nature* that painted a further dire picture of the likelihood of (ongoing) rapid ice melt of the Antarctic shelf, an event that could render coastal cities—San Francisco, New York, Sidney, Shanghai—uninhabitable within two hundred years.[4] In the fall of 2018, the Intergovernmental Panel on Climate Change (IPCC) accordingly released its most alarming report yet on the likely scenario of serious crises occurring as early as 2040—including massive food shortages, persistent fire and drought, and the planetary bleaching and die-off of coral reefs.[5] Climate change is no longer in a distant future, but palpably now. As of this writing, the globe seems to have crossed (yet another) red line, with atmospheric concentrations of CO_2 exceeding a monthly average of 410 ppm (parts per million) for the first time in millions of years—an average last seen during the Pliocene era (some five million years ago), when the sea was on average ten to twenty meters higher than its current levels.[6] All this is causing a kind of "cascade" effect that renders climate change, in the words of David Wallace-Wells, a "horizon of human possibility dramatically dimmed."[7]

How indeed did modern civilization, with its optimistic "progress" that Klee's angel futilely looks backward and down upon, take us here? What is to be done to change course? That may not even be possible at this point, given the growing number of climatologists who believe we have crossed the "nonlinear tipping point."[8] The ethical urgency of the situation has exerted no little pressure on scholars of the humanities who write about the past, posing existential questions about how and why one writes about literature, art, history, and religion in this crisis. For environmental historians and geologists, climate change has instigated a rich debate about whether a wholly new term to designate our entry into a different geological era is needed, an era that is already bringing radical upheavals and changes to life as we know it. What the "Anthropocene"—for this is the first such era where the human leaves an indelible, geological imprint on the fossil record, be it through the supra-addition of radioactive materials or of an anthropogenic increase in carbon—will end up looking like is certainly open to speculation (and the imaginative spaces of science fiction are certainly one valuable location for pursuing that question). The transition periods marking the beginnings and ends of other geological eras marked by rapid and dramatic climate change suggest how radically different life will perforce evolve as it adapts to unstable and fluctuating conditions. Humanity has already ushered in, according to many, a sixth "great wave" of mass species extinction.[9] What does it mean to study literature or religion in even one hundred years, if coastlines will indeed be six feet higher, the oceans blanched by reef die-off? The temptation to trivialize the humanities in scope of the unfolding crisis is great, particularly under the ratcheting pressures of neoliberalism; and yet, more urgently than ever, we need forms of "tentacular thinking," in Donna J. Haraway's words, modes of philosophizing and conceptualizing our entanglements with matter in fundamentally new ways. Haraway riffs off a line drawn from philosopher Isabelle Stengers (who borrowed it from Virginia Woolf): "Think we must. We must think!"[10]

There are many arguments about modernity's ecological predicament that anatomize the "bad thinking" that got us into this pickle in the first place: tracing the root of the problem back to the foundational anthropocentrism of Judeo-Christianity, or to the disenchanting aspects of secularity, or the mind/body dualism that has underwritten much Western philosophy following René Descartes (and later on, Immanuel Kant). Too often these critiques have simplified complex questions in order to assign easy blame, typified by Lynn White's influential contention, in 1967, that Christianity was "the most anthropocentric religion the world has seen," precipitating the environmental crisis as a result.[11] Whatever the cause, our unfolding Anthropocene is inevitably entangled in the complexities generated by what it is we mean by the term modernity, both as an (unfinished) ongoing process and as an historical bracket used to demarcate a particular relationship to the past (though here, too, as Latour is fond of reminding us, we should eschew Whiggish before/after dualisms, for perhaps we were never really modern in the first place).[12] Modernity, imagined or otherwise, has been remarkably persistent in presenting a vision of the world as dead, lifeless, and meaningless—as first prominently diagrammed in Max Weber's account of the disenchantment (*Entzauberung*) of a world vacated by God and transcendent meaning.[13] If the teeming stuff of life was not anchored to any ontotheological coordinates, according to Weber and his followers, the progressive efforts of science and technology were at least capable of mastering it, and "nature," all that was not human, not culture, could be wholly subsumed by human rationality. Nature became both a blank, a void of Divine absence awaiting human inscription (by cartography, the encyclopedia, and other appurtenances of the Enlightenment), and also a bank—a stock of apparently inexhaustible resources awaiting extraction by the ingenuity of commerce (that old Protestant ethic, hard at work). Modernity had replaced religion's superstitious magic with the calculated certitude of math.

Or so the standard story goes. Squarely within the Enlightenment contours of this Weberian anomie, the life and work of the Scandinavian scientist-turned-mystic Emanuel Swedenborg (1688–1772) upsets many of the settled assumptions about modernity's disenchanting telos. This book is about Swedenborg's visionary theology, and how it came to operate as an agent of reenchantment for a group of American authors and artists who are associated with the appearance of an environmental imagination in the nineteenth and early twentieth centuries. Swedenborg, "the translator of nature into thought," as Ralph Waldo Emerson strikingly put it, is a forgotten, vital source for a current of proto-ecological ideas that streamed out of Concord Transcendentalism into the broader, rippling circles of conservation and preservation that mark the emergence of both modern American political environmentalism, and forms of non-institutional "dark green" nature-religion and spirituality.[14] Swedenborg's legacy is also embedded within the ambivalence of the Anthropocene. His career as a leading Scandinavian natural scientist led to formative contributions in geology, crystallography, and mineralogy. The practical implications of this work, summarized in his mechanistic magnum opus, the *Opera Philosophica et Mineralia* (1733–34), led to innovations in smelting technology and the extraction of metal ores from the earth. James Watt, inventor of the steam engine in the 1760s—the machine that veritably generated the Industrial Revolution—was but one of several savants who adapted and used material from Swedenborg's *Opera Philosophica*.[15] Watt's engine has been cited as a harbinger marking the advent of the Anthropocene and our carbon-based civilization. Swedenborg, thus, stands not only as a natural philosopher whose ideas are constitutive to modernity's mechanism, but also as a mystic, the creator of an organicist theosophy that imbued later forms of Romantic ecology that were resistant to the Industrial Revolution. These antinomies within Swedenborg are paradoxes we continue to live with today.

For Emerson, John Muir, Sarah Orne Jewett, and the other writers and painters that will be discussed in this book, Swedenborg's ideas enabled an immanental cosmology that made nature "vibrant," in Jane Bennett's sense of the term, or an enchanted "language of things," as it was coined by Sampson Reed, an early Swedenborgian Romantic. In some cases, to engage with this tutelary language of nature, to attempt to descry and translate it into essays, poems, and paintings that would all adequately "correspond"—a signal term for Swedenborg, as we shall see—between inner and outer, between the fact of feeling in perception and the phenomenal richness of experience outside of ourselves, inculcated early American environmental impulses to conserve and preserve.[16] Flawed as many of these Romantic nature movements now seem in critical hindsight, reifying as they did an arbitrary divide between nature and culture, it nevertheless valuably underscores the symbiotic relationship between the space of aesthetics and the political realm of legal action. The weak governmental response to the present environmental crisis is, if anything, a colossal failure of the imagination—of the imaginary, to be more precise, as a collective social practice.[17] We sorely need not only better international climate legislation, but also more kinds of art and literature that can disrupt and challenge our habits of thinking, and broaden the capacity to reimagine relationships with the natural world. The adaptations of a Swedenborgian "language of things" that are unfolded in this book—and all the strange, queer baggage that Swedenborg brings with him, his angels, trances, and other forms of mystical ecstasy—offer points of departure within an American tradition of nature writing and ecological aesthetics.

Johnny Appleseed's News

In order to situate Swedenborg's place in American ecological imaginaries we can begin with one of his earliest readers and promulgators, a figure invoked today as an icon of early conservation and proto-environmentalism. Toward the end of the eighteenth century, John Chapman, better known by his folk moniker Johnny Appleseed, began his westward peregrination toward the Mississippi through the still-then frontier of the Ohio River valley. A committed pacifist and vegetarian, dressed in motley homespun rags and reputedly walking through the rough woods in his bare feet, Appleseed came to be revered by local Native Americans as a medicine man of great power. As he traversed the Midwest, establishing various landholdings that he subsequently cultivated into orchards, he began his famous distribution of fruit seeds and apple tree seedlings. In addition to leaving his namesake seeds at the homes of various families he would stay with, Chapman occasionally left behind books and pamphlets by and about Swedenborg. "Good news, right fresh from Heaven!" Chapman is said to have announced to the strangers he met on the frontier, as he excitedly shared the pages of Swedenborg's mystical theology with them.[18]

Appleseed's forward-thinking, his planting and reforestation for generations to come, were important early nineteenth-century precursors of modern ecological sustainability, paralleling similar forestry practices of *Nachhaltigkeit* in Germany. While the greening of Appleseed is a much later twentieth-century construct, and his name continues to be instrumentalized for a number of contemporary modern-day environmental projects (such as the Johnny Appleseed Junior Ecology Club), it is rarely recalled how this green archetype of Americana was actively engaged with Swedenborg's visionary teachings. Indeed, Chapman's readings in Swedenborg were responsible for many aspects of his character that lent to his subsequent incorporation into the mythical roots of American environmental thought. His vegetarian principles extended to legendary abstention from the killing of any sentient life, from rattlesnakes to wasps, mosquitoes, and flies. Such anomalous behavior made Chapman the brunt of many frontier jokes, as the historian Robert Price notes, and Chapman acquired early on a folk reputation as the "American St. Francis of Assisi." Chapman's benevolent attitude toward nonhuman nature is not all that surprising, writes Price, when we recall how Chapman gleaned from Swedenborg how "all things in the world exist from a Divine Origin . . . clothed with such forms in nature as enable them to exist there and perform their use and thus corresponding to higher things."[19]

In the earliest visual representation of Chapman known to exist, a drawing from an 1862 history of midwestern pioneers, Chapman stands barefoot in the woods, his left hand cradling a young apple sapling. Tucked into his homespun overalls, a large bulky book peeks above the waistline: seemingly incongruous with the rustic character of everything else in the image that denotes a frontier or wilderness context. This book is probably meant to suggest one of Chapman's beloved Swedenborg volumes, already by this point an indelible part of the growing Appleseed legend. As Michael Pollan puts it, Chapman's "Swedenborgian beliefs must have lit up the whole landscape—the rivers and trees, the bears and wolves and crows, even the mosquitoes—with a divine glow. . . . Chapman's mystical teachings veer about as close to pantheism and nature worship as Christianity has ventured."[20]

Taking its cue from Appleseed's emblematic story, *A Language of Things: Swedenborg and the American Environmental Imagination* is concerned with the presence of Swedenborgian thought within these nineteenth-century conceptualizations of the American environment. How did the mystical theology of Swedenborg, originally written in a dry and reasoned neo-Latin, end up becoming evangelized by this proto-conservationist on the illiterate American frontier, and part of a broader history of the emergence of a different way of looking at nature that coalesced during the nineteenth century?

Appleseed was far from alone in his receptivity to this "good news, right fresh from Heaven." At the same time that Chapman's orchards on the Ohio frontier were flourishing, Swedenborg was beginning to be read by Ralph Waldo Emerson and his circle in Concord, Massachusetts, quickly becoming an instrumental agent for fomenting New England Transcendentalism—Swedenborg was, as Perry Miller put it, "as significant as Coleridge or Carlyle" for that heterogeneous movement's inception.[21] These Transcendentalist energies helped generate, in turn, the philosophical foundations for what became American environmentalism in the latter half of the century, as Lawrence Buell, Roderick Nash, and others have demonstrated.[22] *A Language of Things* excavates this hidden genealogy running from Swedenborg to Emerson, and on into the environmentalist poetics and advocacy projects of later nineteenth-century American writers and artists. By tracing the varied ways that Emerson, the naturalist John Muir, and the regionalist writer Sarah Orne Jewett, as well as the landscape painters George Inness and William Keith, all variously responded to Swedenborgian thought, *A Language of Things* uncovers the complex dynamic that came to exist among the religious, the literary, and the ecological.

These religious dynamics have not been always so clearly understood or taken into full consideration by ecological literary criticism (or ecocriticism, for short) and the standard histories of environmentalism. As John Gatta, Kate Rigby, and others have maintained, ecocritical readings of nineteenth-century literature have tended to replicate certain shibboleths of M. H. Abrams and the older historiographies of Romanticism: that the period evinces "the secularization of inherited theological ideas and ways of thinking," as Abrams influentially claimed.[23] Emerson, Muir, and Jewett all used Swedenborg for projects of reenchantment that mark not so much a return to the religious, in the sense of organized institutions, as much as they refract a postsecular relocation of spiritual authority within naturalized experience. This locus, *A Language of Things* argues, harbored an emergent environmental politics.

A Language of Things is not by any means an attempt to "rescue" Romanticism vis-à-vis Swedenborg from the usual ecocritiques lobbed against it:

that its subjectivity entrenched Cartesian dualism, and all the attendant forms of anthropocentrism and speciesism that go along with it. There's no going back to the aesthetic ideologies of the nineteenth or early twentieth century, as some early environmental literary criticism was fond of implying, as if all we needed for better contemporary green politics was more reading in William Wordsworth or Henry David Thoreau (though such a broadened ecoliteracy certainly couldn't hurt). My claim is rather how a careful parsing of the diffusion and adaptation of Swedenborgian ideas in these earlier American environmental aesthetics can, firstly, restore the formative role of religious experience for generating environmental poetics, a fact that has tended to be neglected, if not repressed, by ecocriticism. Secondly, I hope to illuminate how these Swedenborgian reenchantments of nature anticipate some of the current concerns with immanence and sentience in the so-called "new materialisms" of environmental studies. Just as work in these emerging fields have chosen as significant points of departure critical figures from the eighteenth and nineteenth centuries—ranging from engagements with G. W. Leibniz's monadism, Benedict de Spinoza's *conatus*, to Friedrich Schelling's *Naturphilosophie*—this book aims to bring a set of ideas derived from Swedenborg into dialogue with some of these explicitly posthumanist concerns.[24] Swedenborg matters, this book posits, as his ideas formed part of a habitus of perceiving the more-than-human world that shaped efficacious environmental ethics. This book is an attempt to think critically *with* Swedenborg's eighteenth-century thought and its later transpositions—rather than just historicizing it, and doing yet another study of Swedenborg's influence—much in the same way that Isabelle Stengers has made pragmatic use of Alfred North Whitehead, thinking in concert with his early twentieth-century ideas to read contemporary science and metaphysics.[25]

Reading Swedenborg through the new materialists can be seen as an amplification of an original observation made by the Pragmatist philosopher and semiotician Charles Sanders Peirce. In an 1891 review of William James's *Principles of Psychology,* Peirce tellingly noted how James had been raised within a kind of Swedenborgianism that was "materialism driven deep, and cinched on the inside."[26] Tracing the contours of Swedenborg's life, the movement from natural philosophy into anatomy and physiology, and the final swerve into theophanic mysticism, *A Language of Things* will seek to locate such deep, cinched places, the Swedenborgian "materialism driven deep" that Peirce noted in James. These places may be where "consciousness cease[s] to be a light cast upon objects in order to become a pure phosphorescence of things in themselves," as Gilles Deleuze has described his own attempt to get beyond normative epistemological paradigms.[27] Such phosphorescence of things, for Swedenborg (unlike Deleuze), made matter glow and become numinous, in Rudolf Otto's troping on that word: a core (religious) feeling about the encounter with a decentering otherness.

This numinous potential of Swedenborg's writings, marginalized and contraband within Sweden during his lifetime, later gave shape to alternative models of consciousness that flourished across nineteenth-century America, as both John Haller and Alfred Gabay—historians of medicine—have fully explored.[28] This is signaled perhaps by an odd moment in Emerson's journal, in 1854, where Emerson reflects on the pervasive currents of mesmerism and spiritualism that were swirling around him (forms of popular esotericism that, like phrenology and homeopathy, frequently cited Swedenborg's writings as an authenticating, guiding light). "The Age," Emerson dryly writes, "is Swedenborg's. . . . [It has] embrace[d] the subjective philosophy of the Saxons, that the soul makes its own world."[29] Emerson's phrasing, "The Age is Swedenborg's," became F. O. Matthiessen's quip that his formative *American Renaissance* (1941)—a foundational text for the inception of American studies as a disciplinary field—could have been alternatively titled "The Age of Swedenborg," had it been written under different inflections.[30] Several scholars could be said to have developed the framework intimated by Matthiessen, from Leigh Eric Schmidt's exploration of the role of Swedenborgian ideas within American post-Enlightenment mind–body conceptual paradigms, to Catherine Albanese's influential tracing of Swedenborg's impact on what she calls "American nature-religion": a popular, symbolic center that took shape throughout nineteenth-century American thought, united by a tendency to sacrilize nonhuman, natural space.[31] Nevertheless, in spite of such citations by these prominent scholars of American religious history, both literary ecocriticism and environmental histories have paid scant attention to Swedenborg, and he remains, by and large, a matter relegated to footnotes, an eighteenth-century stumper question for Double Jeopardy.[32]

No book has yet answered Emerson's tantalizing provocation in his *Representative Men* (1850), that he "sometimes thought" how someone "would render the greatest service to modern criticism, who should draw the line of relation that subsists from Shakspeare [*sic*] to Swedenborg"—in other words, to account for the convergence between the religious, or mystical, typified by Swedenborg, and the literary as represented by Shakespeare (Emerson's signal poet in *Representative Men*). *A Language of Things* attempts to draw Emerson's line of symbolic language, if not back to Shakespeare, then forward into the emergence of American environmental imaginaries.

Lines of Relation

Due in part to the general unfamiliarity with Swedenborg's work, in the first two chapters I will provide an overview of Swedenborg in his eighteenth-century context, (re)reading his science and theology from an ecocritical perspective. The story of Swedenborg's life has been told from the vantage point of the history of ideas (Martin Lamm, Inge Jonsson), the development of Protestant theology (Friedemann Stengel), or esotericism and espionage (Marsha Keith Schuchard). Chapter 1, "Planetary Pictures" will attempt a first: to see the evolution of Swedenborg's thought through a specifically green lens. Attention will be given to both those works that became critical interlocutors for eighteenth-century natural philosophy, such as his *Principia* and the *Regnum Animale* (The soul's domain), but also lesser-known pieces that foreshadow his later tilt into visionary theosophy. With this biographical context in place, the second chapter, "Psychogeographies of Heaven and Hell," turns toward a set of ideas that thread through Swedenborg's mature theology, which become, later, so significant for American environmental imaginaries: first, his doctrine of "correspondences" that transformed nature into a kind of personalized, sacred language; and secondly, his idea of "influx," a belief that nature was shaped by a ceaseless flow of divine emanations, a vitalist modulation of Neoplatonism that broke down mechanistic models of the universe. Swedenborg's correspondence and influx theories will be examined in light of several concepts that have been generative for the new materialisms, including the sign theories of Peirce, which remain an important touchstone for biosemiotics, as well as Gilles Deleuze and Félix Guattari's "Body-without-Organs"—the BwO—from their *Thousand Plateaus.*

Like the new materialisms that have drawn on Deleuze to critique the atomized, singular, and self-contained subject, at the center of Swedenborg's mature theology was an unsettling deconstruction of the individual self: a conviction that the self was, in essence, what Wallace Stevens would later term "a necessary fiction," a no-thing in and of itself. Daisetsu Teitaro (D. T.) Suzuki (1870–1966) found Swedenborg's ontology on this point to come closer to nondualistic Buddhism than to anything in the Judeo-Christian tradition from which Swedenborg had emerged, warranting a view of Swedenborg as a translocative "Buddha of the North" for Europe and the West.[33] This "buddhistic" strand of Swedenborg parallels the new materialism's preoccupations with Buddhism as an alternative locus for environmental ontology (if not an inchoate form of ecological spirituality).[34] For the authors under discussion in this book (who were so often also engaging with Asian religious traditions),

Swedenborg could present an indigenous, "home-grown," nondualistic form of theosophy within the fabric of Judeo-Christianity—a fact overlooked by most ecocriticism. In spite of posthumanist ecocriticism often claiming to go beyond the Enlightenment—that "Enlightenment rationality is not, as it were, rational enough, because it stops short of applying its own protocols and commitments to itself," as Cary Wolfe nicely puts it—in its contorted attempts to "strip nature of its metaphysical halo" (Hannes Bergthaler's phrasing), or through its polemics against any kind of poststructuralist "return to religion" (in Quentin Meillassoux's speculative realism), some of the new materialists have ended up reiterating the Enlightenment's self-given assumptions about the homologies between modernity and the secular.[35] Swedenborg's place within American nature aesthetics, by contrast, reveals troubling areas where the Romantic posthuman is thickly entangled with the postsecular.

The title of the book, *A Language of Things,* is paraphrased from a Swedenborgian text by Sampson Reed that was first published in 1826, and is regarded today as one of the seminal manifestos for American Romanticism. Reed's little book notably features one of the first calls for a free-verse poetics, as will be discussed. Emerson read Reed's *Growth of the Mind* repeatedly over a fifty-year period, and enthusiastically gave copies of the book to Thomas Carlyle, Henry David Thoreau, and John Muir, among others. Reed writes, transposing Swedenborg's doctrine of correspondences into more poetic language, how "there is a language, not of words, but of things. When this language shall have been made apparent, that which is human will have answered its end; and being as it were resolved into its original elements, it will lose itself in nature."[36] Here, Reed articulates a desire for the logos to "lose" its humanness, for words to mesh with material things. This fantasy of a linguistic return—"Adam and Eve knew no language but their garden," as Reed parabolically writes—became strongly amplified in subsequent American Transcendentalism.

Chapter 3, "Radical Correspondence," examines the Transcendentalist activity of making "things into words," the specific ways Swedenborg's correspondence and influx came to imbue the environmental imagination of Ralph Waldo Emerson.[37] Bolstered, in part, by a predilection for viewing landscape symbolically as Scripture, a habit traceable back to Puritan traditions of typology, chapter 3 argues that Swedenborg did not simply (and problematically) aid Emerson in transforming nature into an abstract, anthropocentric sign, but that Swedenborg's language of things aided Emerson in *feeling* what he called a "ray of relation," that "shot from every other being" to him.[38] Correspondence became to correspond with the other-than-human, to commune with what was held in common; "because of this radical correspondence between visible things and human thoughts, savages, who have only what is necessary, converse in figures."[39]

This third chapter on Emerson sets in motion the two-pronged approach for all the subjects in the book. A historicist claim for the overlooked significance of Swedenborg will be supported by due attention to marginalia, material in archives, letters, journals, and the like, at the same time that a consideration of these adaptations of Swedenborgian theology, their translation into aesthetics or poetics, resonates with some of new materialist and posthumanist strands of ecocriticism mapped out in the second chapter on Swedenborg. As Swedenborg was by no means read in a vacuum by any of the figures in this study, the broader contexts of their cosmopolitan appropriations will be important to acknowledge. Indeed, other eighteenth- and early nineteenth-century thinkers, such as Alexander von Humboldt, certainly left clearer ecological traces on the environmental imaginations of Emerson and other Transcendentalists, as several recent books have so artfully demonstrated.[40] The story of Swedenborg's environmentalism, however, quite unlike Humboldt's, has not been told, and the contribution of Swedenborg's language of things was an important ingredient of a perennialist eclecticism informing an American habit of turning nature into a sacred text, "hieroglyphics written with sunbeams," as John Muir cheerfully describes his beloved California mountains.

Chapter 4, "Herald of a New Gospel," addresses the circle that developed around John Muir, the concurrent birth of the conservation and preservation movements, and the ultimate establishment of the Sierra Club in 1892 (which initially met in the art studio of the Swedenborgian landscape painter William Keith). Not coincidentally, Muir's formative encounter with Emerson in 1871 concluded with Emerson giving the young proto-conservationist a copy of Reed's *Observations*, with its oracular pronouncements about correspondence and the language of things. Muir substantially annotated the text, and other subsequent works by Swedenborg (marginalia that has been largely ignored by scholars investigating the religious dimensions of Muir's project). The chapter concludes with a consideration of Muir's *Stickeen* (1897), a memoir about Muir's friendship with a Native American dog that stages an ethical relation with the other-than-human world. The chapter title comes from Muir's draft notes to *Stickeen*, where Muir described his preternatural canine friend as "a herald of a new Gospel, a Swedenborg." An alternative chapter titling here, thus, might have sounded a bit like the start of a joke that riffs off of Deleuze and Guattari's "Becoming Intense, Becoming Animal" essay in *A Thousand Plateaus:* (just) how is John Muir's dog like a scientist and mystic (or, conversely, how is a Swedenborg like a dog)—what animal-human becomings are unfolded in Muir's dog story if we pay close attention to his Swedenborgian marginalia?

Muir's wilderness machismo that abetted early conservation efforts was deeply problematic, however, entangled as it was in American expansionist policies and a discourse of aesthetic sanctity that was the province of subject positions often exclusively white and male. Muir's encounter with Swedenborgian ideas took place in a milieu of nature aesthetics and environmental conservation where his painter friend Keith and the reclusive architectural guru and Swedenborgian minister Joseph Worcester figured prominently. The unique church structure that Keith and Worcester helped build, in 1895, a so-called "garden church" that functioned as a veritable temple to wilderness spirituality, epitomizes many of the paradoxes of the Arts and Crafts "back to nature" movement that the structure helped launch. The exquisitely beautiful space on the one hand de-anthropocentrizes the ostensible norms of Christian ecclesiastical architecture, often in radical and innovative ways. Yet on the other, the materials in the building depended on a circuitry of wealth and extractive capitalism, from strip-mining to clear-cutting forests, for the materials to have been incorporated into the building.

Chapter 5, "Homes for Herons," explores a very different kind of ecological inflection of Swedenborg within the regionalist fiction of Sarah Orne Jewett, and a body of parallel paintings created by the artist George Inness, a Swedenborgian paragon of American Tonalism. Claimed as a significant feminist author, the subject of more recent queer readings that have illuminated networks of same-sex desire in her texts, Jewett—more or less contemporaneous with Muir—has been an important site for ecocritical reappraisals of nineteenth-century Regionalism. Almost all of these readings, however, have proceeded without giving any attention to Jewett's sustained interest and engagement with Swedenborg. Jewett's most well known environmental text, the short story "A White Heron" from the eponymous collection (1886), will be situated into a dialogue with contemporaneous paintings of herons made by Inness—who, like his friend William Keith, attempted to derive an aesthetic theory of representation from Swedenborgian theology. I argue that Inness's paintings of herons in their habitat and Jewett's titular story are inextricable from a groundswell of ornithological conservation that resulted in the appearance of Audubon societies during the last decade of the nineteenth century, and the ultimate passage of the Lacey Act, in 1900.[41] Inness and Jewett's coterminous turn toward investing birdlife with a numinous, spiritualized otherness is mutually shaped by their engagements with Swedenborgian thought, and demonstrates how the Swedenborgian language of things need not always be aligned with the frontier telos of the wild that Muir embodies (and is so often critiqued for); that Jewett, in particular, suggests a different, queer location for tracing Swedenborg's contribution to an American environmental imaginary.

The book concludes with a brief return to the opening figure of Johnny Appleseed, using Chapman's enduring presence in American environmental imagination as a means for offering some broader speculations on the place of religion in current ecocritical debates about the Anthropocene. Appleseed's alien otherness on the frontier, his way of viewing landscape as alit "with divine glow," as Pollan puts it, and his profound reverence for all sentient life, further illuminate two signal terms that are worth glossing now, as they will recur throughout this project: firstly, the *numinous,* a concept of the religious rooted in the work of Rudolf Otto, as already noted, and, secondly, the *other-than-human.* Otto's numinousness was at its core an a priori category of felt experience that resists schemas of rationalization or systematic dogma: it was a "*mysterium tremendum et fascinans,*" a "mystery before which one both trembles and is fascinated," an experience of transcendence beyond the self, yet embedded in the sensual phenomenology of the body.[42] As Peter Schneck astutely notes, "Otto's concept of numinous experience as a desecularized version of the sublime links his idea of the holy to the fate of aesthetic experience and art within the larger framework processes of secularization."[43] For *A Language of Things,* the "numinous" invokes this nondogmatic sense of the holy that is entangled in a broader problem of the status of art and nature within the dynamics (and discontents) of secularization. It is a further useful idea to deploy, given its long traction within ecocriticism and environmental histories concerned with religion. The numinous in nature is akin to other conceptualizations that have endowed matter with a "shimmer" enmeshing the human and the other-than-human (Deborah Bird Rose), or a kind of "vibrancy" of things that is vitalistic (Jane Bennett).[44] Otto's nomenclature has been retained here for its origin within a Protestant milieu, the same theological soil churned by the Pietist currents that earlier gave shape to Swedenborg in the eighteenth century. The spiritual transcendence inhering to the ideas of Otto and Swedenborg make them different, yet nevertheless still related, to these other attempts to move matter out of Western Cartesian frameworks into other, more animate sorts of spaces of thinking.

Secondly, with the "other-than-human" (or as it has also been sometimes articulated, the "more-than-human"), I mean to indicate the work undertaken by anthropologists and other social scientists on the ascription of agency and ontological being—the constitutive elements of personhood, in other words—onto things beyond the human: to rivers, mountains, and dogs, for example. Though landmark studies by David Hallowell and Graham Harvey ostensibly locate the origins of the "other-than-human" in their approaches to non-Western, indigenous cultures and what we would call various forms of animism, *A Language of Things* hopes to demonstrate the perhaps surprising capacity of the post-Enlightenment Western canon for providing vitalizing counterexamples to modernity's dead-end, disenchanted nature.[45] From Emerson's vegetables in his *Nature* essay—"They nod to me, and I to them"—to Muir's dog Stickeen, to Inness's and Jewett's glowing, ephemeral herons, these beings are endowed with varying degrees of an other-than-human personhood.

Swedenborg's Garden

What kind of a view on the American environmental imagination might emerge after reading this disparate sequence of authors and artists, and their varied uses of Swedenborg? Several years after undergoing a profound existential crisis, and shifting away from the natural sciences into exegetical theology, Swedenborg built an elaborate garden surrounding his home on the island of Södermalm, in southern Stockholm. It was divided into four sections, with a central avenue of lime trees running down the center. At the back of one section, facing the main house and a pavilion, stood a small summerhouse, painted in yellow—a traditional Swedish *lusthus*—with windows that opened to the flowerbeds and vegetable gardens around it. It was here that Swedenborg came write and record many of his famous visionary texts.[46] Parts of the garden that Swedenborg could look out on contained exotic and rare specimens from the North American continent, such as maize, dogwoods, mulberries, buttonwoods, and American beech trees: an appropriate view, given the fertile soil that Swedenborg's ideas would come to find in the United States in the subsequent century (much more receptive ground, as we shall see, than the cold climate of Swedish Lutheranism). Across the street from Swedenborg's property, a steep embankment led to cliffs that still command a fantastic panorama of Stockholm. To the left and below glittered the freshwater Lake Mälaren, and to the right, the Baltic archipelago began, with the spires of the cathedrals and churches, the Renaissance buildings of the Royal Palace and House of Nobles, rising up from the old medieval city directly across the way on the island of Gamla Stan. Swedenborg's garden figured on the border between eighteenth-century culture and nature, the growing cosmopolitan districts of Stockholm on the one hand, and the wilder spaces of the outlying archipelago and countryside on the other.

Swedenborg's garden also functioned as a space of reenchantment, containing horizons of contemplation that were available both internally—in the detailed layout of its botanical collections, in the growing garden seeds—and externally, along the blurred line of sky and sea that circled Södermalm island. Sounding much like his cousin and scientific colleague Carolus Linnaeus (1707–1778), Swedenborg would come to wonder in his *Divine Providence* (1764), a work likely penned at least in part in the *lusthus,* how "there is an image of what is infinite and eternal in the way everything bears fruit and multiplies in the reproductive ability of seeds."[47] Swedenborg continues in a Linnean vein: "Just look at a fruit tree. See how it is first born from a tiny seed as a delicate sprout, how this gradually develops into a trunk that sends out branches, how these are covered with leaves, and how it then produces flowers, bears fruit, and sets new seeds in the fruit that provide for its endless future." Like Linnaeus, the Scandinavian father of modern botany and ecology, Swedenborg frequently tropes on an affect of wonder in such passages—to contemplate a "silkworm or a bee" (his garden had, for a period of time, American mulberries with silkworms) could but reveal how a teleological "force has been instilled into everything that has been created," something that should but "stun" and "amaze" the mind.[48] Against the prospect of a mechanized, dead material world that the first half of his Enlightenment career had been exclusively devoted to cataloguing, Swedenborg's mystical writings opened new conduits of feeling for an other-than-human vitalism that sprouted and grew in the garden around him. Swedenborg's garden offered what ecocritics have called "a story-laden mode of reenchantment" that "demands an ethics of relation, entanglement, and wonder."[49] As a space for visionary experience—it provides the mise-en-scène for several of Swedenborg's infamous accounts of conversations with angels and devils—the garden certainly contained "effects dramatic and strange," as Bennett might put it, in her own descriptive attachments to the "vibrant matter" of trash and detritus that she finds in a Baltimore street gutter. And yet, Swedenborg's garden certainly inspired in him "a greater sense of the extent to which all bodies are kin in the sense of [being] inextricably enmeshed in a dense network of relations."[50]

What do we do with the inevitable supernatural bodies, the spirits and angels entangled in some of these relations? "On one occasion," Swedenborg dryly records, probably from his desk in the *lusthus,* "A satan was allowed to come up from hell. A woman came with him. He came to the house where I was. When I saw them I closed the window, but I did talk to them through the glass."[51] Such matter-of-fact detailing of the paranormal continues to trouble Swedenborg's readers. In a long essay on the painter George Inness in the *New York Review of Books*, for example, the art critic Andrew Butterfield acknowledges Swedenborg's nineteenth-century popularity, but that today, "he is routinely dismissed as a crank."[52] In Jose Luis Borges's beautiful essay on Swedenborg from 1975, he notes an "enormous wall of bronze" of incredulity and skepticism that he could feel welling up in his readers against Swedenborg's claims to speak with the dead. Yet, Borges protests, sounding a little like Bruno Latour ("we have never been modern!"), "We believe in Ezekiel because he is exalted by the remoteness of time and space; we believe St. John of the Cross because he is an integral part of Spanish literature; but we do not believe William Blake, Swedenborg's rebellious disciple, nor his still-recent master. What was the exact date when true visions ended and were replaced by apocryphal ones?"[53]

One date might be 1766, the year Immanuel Kant published his *Träume eines Geistersehers* (Dreams of a spirit seer), an ambivalent polemic against Swedenborg that had been triggered by Kant's initial curiosity about Swedenborg's reputed preternatural capabilities. Swedenborg's ultimate cardinal sin, Kant opined, was thinking that spiritual *noumenon* could be empirically known and verified by the senses.[54] There might indeed be a spiritual world out there with intelligible angels, but there was no way of grounding this as a rationally provable phenomenon. Satan simply does not stroll up to your summerhouse, requiring you to shut the window against a stink of sulfur. Kant's scathing book guaranteed that Swedenborg's reputation took a nose-dive among Lutheran readers across Germany and Scandinavia, even as the book's satirical tone seemed to mask the ways that Kant's system of transcendental philosophy closely mimicked Swedenborg's own.[55]

By contrast, for Swedenborg (as it was for Blake), heaven really was "in a wildflower / and eternity in a grain of sand"—the concrete and minuscule particulars of the physical world were ways of knowing the spiritual essences of things, the *ding-an-sich*.[56] The last two hundred and fifty years of Western philosophy have largely followed Kant, not Swedenborg or Blake, on this point, and duly banished "religion" or "spirit" to a private realm of feeling while buttressing the emergent public spheres of democratic institutions as largely secular zones. The planetary crisis of our Anthropocene may be unsettling some of these assumptions, however, especially given the spectacular failure of the secular regimes of international environmental law; it is a dawning new age, if anything, of the return of the uncanny and the monstrous, of imagining the planet as a giant entity (James Lovelock's Gaia) or a multitentacled kind of creature (Donna Haraway's Cthulu).[57] As outlandish and otherworldly as the paranormal in Swedenborg initially seems, his angels and spirits may be metaphorical kin to some of the planetary weirdness we now find ourselves in. What could be stranger than what the Anthropocene is demanding of us: a present that is defined by imagining a desiccated future whereby some disembodied geologist looks back in time and traces the strata of our own anthropogenic extinction?[58] This is the "end of history," although not in the blithe way Francis Fukayama once predicted, but as the temporal crossroads where humanity's very geological force now obliterates our self-determination and agency.[59]

The planetary was present in Swedenborg's garden in a very real way, in how its layout microcosmically reflected the known botanical world. Besides the exotic American specimens, Swedenborg went through great lengths to acquire the rarest, most obscure plant specimens from Africa, Asia, and the Americas, and accordingly arranged them around geographic distribution.[60] For the more delicate, tropical species, such as orange and lemon trees, Swedenborg converted the second floor of his modest house into a large conservatory, with glass skylights in the roof and openings to the ground floor so that the fragrant plants—and the sound of singing birds from a *volière* (a large cage of birds)—poured into the sleeping, living, and writing chambers below. Thus, whether he was writing in his summerhouse outside in the warm months, or indoors below the conservatory during the winter, Swedenborg was surrounded by a kind of terrestrial heaven of his own making. No surprise, then, that the space and imagery of gardens occupies such a prominent location in the subsequent theology. He writes in *Heaven and Hell* (1758), in a passage that Emerson later copied into his journal:

> [Angels] live in gardens where you can see flower beds and lawns beautifully marked off, surrounded by rows of trees with arcades and promenades. The trees and flowers change from day to day. Looking at all this brings pleasure to their minds generally, and the specific changes make it constantly new. Further, since all this corresponds to divine qualities, and since these people are drawn to their knowledge of correspondences, they are constantly being filled with new insights and thereby having their spiritual rational faculty perfected. They enjoy these pleasures because gardens, flowerbeds, lawns, and trees correspond to information, insights, and the intelligence that ensues.[61]

By learning from the text of beauty around them, Swedenborg's angels, like Henry David Thoreau, live out Emerson's declaration that nature should be "a mute gospel . . . a sacred emblem from the first furrow of spring to the last stack which the snow of winter overtakes in the fields."[62] In Thoreau's *Walden* and Swedenborg's heaven, close observation of natural phenomena leads to spiritual enlightenment.

Before we enter Swedenborg's garden and observe the strange American transplants that came to grow there—starting, perhaps, with a token apple seedling from Johnny Appleseed—it is necessary to take note of some specimens that sprout on the wayside, which this book will not be addressing in spite of their flourishing in the Swedenborgian soil. A seemingly conspicuous omission will be Walt Whitman, who, following Emerson, took and read Swedenborg quite seriously, and even sketched out a longer project to write an essay about Swedenborg in 1857 or 1858. According to David Reynolds and other Whitman scholars, we can locate definitive traces of Swedenborgian correspondence theory behind Whitman's poetic spiritualization of the natural world; Anders Hallengren goes even further to posit that Swedenborg might provide the specific key that unlocks Whitman's enigmatic tropes around the "leaves of grass."[63] Swedenborg's coupling of erotic with divine love further greatly appealed to Whitman's sensibilities: "I think Swedenborg was right when he said there was a close connection—a very close connection—between the state we call religious ecstasy and the desire to copulate," as he told his confidante Horace Traubel (certainly with a wink or a twinkling grin)—"I find it confirmed in all my experience."[64]

Whitman's enthusiasm for Swedenborg, however, must be seen as subsumed within his voracious appetite for all kinds of esoterica and non-institutional forms of spirituality, such as phrenology and mesmerism. Swedenborg was, in other words, but one of many streams affecting Whitman's ecopoetics. For the parameters of this book's contribution to the history of ideas, figures with more substantial—if not singular—investment in Swedenborg, as demonstrable through their marginalia, letters, and the like, have been weighted with greater importance.

Similarly, Henry David Thoreau is not singularly addressed here other than as part of the orbit of Emerson's broad circumference of influence. In spite of some notable attempts to locate a "Swedenborgian spirit" in Thoreau's writings, it is hard to get past Thoreau's own blunt assessment that he "had no practical use in Swedenborg."[65] The intellectual biographies of Thoreau by Laura Dassow Walls and Robert Richardson have nary a word regarding Emerson's representative mystic, and when it came to reading eighteenth-century Swedes, Thoreau clearly preferred his Linnaeus.[66] Thoreau may not have read much, if any, of Swedenborg's writings themselves, but he did have several notable encounters with Swedenborgian concepts in the work of Sampson Reed and James John Garth Wilkinson, the important British nineteenth-century translator of Swedenborg, and close friend of Emerson's. Some of these indirect ripples of reading—perhaps even affecting the surface of Thoreau's *Walden,* as we shall see—will be discussed in regards to Emerson, but it is insufficient grounds to warrant a separate chapter on the so-called "patron saint" of American environmentalism.[67]

F. O. Matthiessen's canon-forming *American Renaissance* that, as noted, somewhat jocularly suggested an alternative title as "The Age of Swedenborg," features key chapters on Emerson, Thoreau, Whitman, Melville, and Hawthorne; originally, Matthiessen had planned to include in the volume one other signal figure who also made significant recourse to Swedenborg: Edgar Allan Poe. The great American representative of the gothic is one of the very few nineteenth-century American writers who seems to have fully digested Swedenborg's scientific cosmogony, with Poe weaving insightful references to Swedenborg's nebular hypothesis from the *Principia* into his prose-poem *Eureka* (the most important thing, Poe claimed, that he ever wrote).[68] In addition to *Eureka,* Poe makes scattered references to Swedenborg throughout his fiction, most significantly placing a copy of Swedenborg's otherworldly *Heaven and Hell* in the library of the eponymous House of Usher. Yet again from the perspective of the history of ideas, this book is not invested in broadly surveying just any fuzzy and vague channels of influence, but in specific instances whereby the aestheticization of Swedenborgian theology harbored implicit or explicit environmental affect. For this present study, then, Roderick Usher's dreary page-turning of *Heaven and Hell* doesn't carry the same measure of significance as, say, Muir's reading of the same text, and the later contribution of a Swedenborgian language of things to the cultural contexts that shaped the appearance of the Sierra Club.

When the following chapters are taken all together, like the expansive vistas of sea and city available from the cliffs of Södermalm adjacent to Swedenborg's garden, *A Language of Things* hopes to provide an equally compelling perspective on some of the dynamics between nature and culture that came to unfold across nineteenth-century America. Ideally, the chapters will prompt further reflection on the horizons of literary form: on the specific ways that a theological "language of things" became transposed into Emerson's essays, Muir's wilderness prose, Jewett's Regionalist fiction, and the experimental landscapes painted by Keith and Inness. These varieties of form might be seen as precursors in the American "game" of language and sign-making that came to be known as Pragmatism.[69] Through their respective struggles to translate a numinous nature into the appropriate words, the figures in this study sought to "bring out of each word its practical cash-value, [to] set it at work within the stream of experience," as William James puts it in his well-known definition of "What Is Pragmatism." As James further writes, such linguistic—and aesthetic—translation "appears less as a solution, then, than as a program for more work, and more particularly as an indication of the ways in which existing realities may be changed."[70]

Setting these Swedenborgian refractions out of Emerson, Muir, and Jewett to play in the streams of our twenty-first-century experience, what of our existing realities might be changed, or reimagined? This is the perhaps unfashionable question that *A Language of Things* hazards. Certainly, reading Emerson on correspondences in nature is not going to provide a better means for addressing the complexities of anthropogenic climate change: only strong intergovernmental regulation and sophisticated scientific models can move toward that with any kind of adequate parity. But by excavating some of these contours of environmental imagination in nineteenth-century America—the patterned ways and habits of perceiving natural space as something sacred that grew out of Romantic reenchantment—we might open new horizons of thought in the present, alternative ways we can translate the language of things into the varied tongues of the human.

1
Planetary Pictures

> In the Anthropocene, both climate science and biology seem to bring back spirits, once thought to have been killed by secular thought, back to life.
>
> —Nils Bubandt

The foundation of Swedenborg's thought begins in the materialistic mechanism of the early Enlightenment, but ends with an adumbration of Romanticism—one of his most ardent and enthusiastic early readers was the London artist and poet William Blake, a figure who typifies Romantic counter-Enlightenment tendencies. Identifying the continuities, trying to understand the breaks and paradigm shifts in Swedenborg's prolific output—so much so, that at some forty-two thousand pages, his manuscripts are one of the world's largest single-author manuscript collections from the eighteenth century—has presented no small challenge to several generations of scholars. To put it in slightly reductive terms, how does one account for Swedenborg's movement from the mechanism of the mine into the mysticism of his garden? At the very least, Swedenborg constitutes, as Inge Jonsson puts it, an overlooked "link between seventeenth-century rationalism and the natural philosophy of Romanticism."[1]

Swedenborg's turn toward vitalistic and immanental theologies, his valuation of intuition and will as primary, over and above the faculties of reason, emerged from a life-long negotiation with the Rationalist tradition that had viewed nature as mechanized dead matter, a quantifiable universe of stuff that could be mastered through the Enlightenment's holy trinity of reason, mathematics, and experience. Only these three things, Swedenborg had proclaimed at the start of his great mineralogical treatise, *The Principia* (1733), were necessary in order to "pierce" to the heart of nature, for "the whole world itself is a pure system of mechanism."[2]

This chapter is primarily concerned with surveying the intellectual and scientific contexts that gave shape to Swedenborg's earlier mechanistic outlook, while emphasizing those countervailing green tendencies that came later to germinate American environmental imaginaries. The chapter title, "Planetary Pictures," is borrowed from James John Garth Wilkinson's seminal 1849 biography of Swedenborg that was read and discussed by Ralph Waldo Emerson and other Transcendentalists. Wilkinson's long first chapter is an attempt to synthesize Swedenborg's multifarious scientific accomplishments, and then square them with the later mystical theology. Wilkinson's book represents one of the first significant efforts to integrate these two halves of Swedenborg's life, the scientific and the theological, and to see them as part of a single continuum.[3] The biography is shaped, in part, by an 1840s radical milieu of reform, science, and revolution. Wilkinson was closely associated with various Fourierist utopian endeavors, and had spent time in Paris during the 1848 barricades, in the midst of his writing and research for the Swedenborg biography—a project that was being bankrolled by the American Henry James Sr. (father of William and Henry, and close friend of Emerson's).[4] In his "Philosopher of Nature" chapter, Wilkinson articulates a desire for a grand theory that could somehow account for the apparently interconnected turmoil of his day, in an architectonic manner akin to Swedenborg's own system-building. "The unity of the world," writes Wilkinson, "is beginning to be recognized as the basis of teaching; the universality of phenomena as the explanatory statement of single facts. The sweep of the ocean currents is seen by the child as part of a *planetary picture*. The fortunes of each trade are found to be regulated by the whole mundane society. Private medicine resolves itself into the question of public healing. And so forth."[5]

Karl Marx and Friedrich Engels had just published their *Communist Manifesto* the year prior, and only seven years later, Ernst Haeckel would introduce to the world the concept of ecology in his *Morphologie der Pflanzen* (Morphology of plants) (1866), fulfilling Wilkinson's wish for a "planetary picture" that could encompass natural and cultural systems as a single unity: the study of the relations within a single household (the λόγος, logos—study—of an οικος, oikos—a dwelling containing different inhabitants, in different sorts of relation with one another). Wilkinson's "planetary picture" also parallels Alexander von Humboldt's *Kosmos* project, a concept beginning to permeate the Anglo-Transcendentalist circles he was part of (Emerson had read the first volumes of Humboldt's *Kosmos* by the late 1840s).[6]

Like Humboldt and other figures associated with an older *Naturphilosophie* tradition that streamed through the aftermath of the Enlightenment in Germany, Swedenborg's proto-Romanticism anticipates some of Haeckel's ecological holism in very general sorts of ways. This chapter and the next, which will turn more explicitly to Swedenborg's correspondence, influx, and micro-and-macro-cosmic analogies, will (re)evaluate Swedenborg's ideas from an explicitly ecocritical perspective, using a set of questions and ideas taken from some of the new materialist and posthumanist turns in environmental thought. Such a "greening" of Swedenborg has not been attempted before, and I will be making necessary recourse to prior studies that have embedded Swedenborg within the history of science and cognition (David Dunèr), post-Enlightenment Protestant theology (Friedemann Stengel), Western esotericism (Wouter Hanegraaff), and world literature (Inge Jonsson, Lars Bergquist, and Olof Lagerkranz). Although posthumanism and the new materialisms are contested terms and inchoate fields, with a variety of differing voices and claims, for any study that takes up a Christian mystic such as Swedenborg, their hostility to the ontotheological presents an immediate challenge: be it an explicit aim to "strip nature of its metaphysical halo," or to divest matter of any "metaphysics of presence" that might come creeping in through the backdoor of critical theory (à la the negative dialectics of Theodor Adorno, or forms of political theologies).[7] As will become evident, Swedenborg's ideas have a natural affinity with those strands of posthumanism and the new materialisms that have developed out of contact with the ideas of Alfred North Whitehead, whose "process" thought continues to generate important environmental theologies, and whose own magnum opus, *Process and Reality* (1929), concludes with an argument for the dynamic exchange of love between God and the world.[8] While these other discrepancies between Swedenborg and the new materialisms must be carefully parsed out—and they have, hopefully, created productive tensions for this book's methodology and approach to its historical subjects—the core of my reading of Swedenborg hinges around two observations that are fairly straightforward, and that further provide useful points of entry into the next section that will provide a summary of Swedenborg's scientific development.

The first observation is that Swedenborg, after his great crisis and turn from natural philosophy to theology, came to feel that life was a flow, a constant sequencing of what Karen Barad has termed "intra-activity."[9] Things (including the human) were no longer definable as singular, stable entities of contained agency, but were caught up in dynamic, ceaseless relations of "influx" and "efflux" into one another, and—for Swedenborg—with the Divine. Swedenborg would have fully agreed with Barad's insistence that "nothing exists that precedes relations." At the macro-level here, Swedenborg's vision of the cosmos became a singular flow of spirits into an enormous human figure, the *Maximus Homo,* or "greatest [biggest] human." This would seem initially to be the crudest of all anthropocentric conceits—that the universe looks, literally, like a giant man—and yet, Swedenborg's *Maximus Homo* is not that far away from a Deleuzian notion of the "assemblage," and its felt embodiment in Swedenborg's own person caused unsettling, displacing effects to Swedenborg's sense of self-identity, further loosening his subjectivity and agency. Swedenborg's written records of his *feeling* this cosmic connection in his body offer ways to read the *Maximus Homo* through another figure that will be drawn from Gilles Deleuze and Félix Guattari, that of their "body without organs," or BwO.

The second observation is even more generic and simple, and that is to underscore how the vibrating source for the flow of matter and life in Swedenborg's systematic theology was, in essence, (Divine) love, an emanating love that caused forms to arise as it united itself with wisdom (or truth). "A kelson of the creation is love," as Walt Whitman poetically puts it.[10] This recurrent key sounds throughout Swedenborg's theosophy, and it shows him in a sustained dialogue with Christian theological traditions, from Augustine onward. "Love" was not just (or only) some abstract principle of creation: it was the key to accessing, for knowing reality. With this overturning of his earlier Enlightenment episteme, where he had formerly championed "reason, geometry, and experience," Swedenborg personifies a central contention made by the Whiteheadian philosopher Graham Harman, a leading exponent of the so-called "speculative realism" that has had a significant impact on posthumanist ecocriticism. "The real is something that cannot be known, only loved," Harman claims, typifying speculative realism's attempt to shift philosophy from epistemology—what can be known—to ontology—questions of being. True philosophy is thus "a *love* of wisdom that makes no claim to be an actual wisdom."[11] The preeminence of love—and its coordinates of wonder, beauty, and ecstasy—that are found in Swedenborg's later work is all the more surprising when we consider the aridity of not only the Rationalist philosophy that he came to contemn, but the pervasive Lutheran theologies he was surrounded by as a young man, where love was secondary to having an intellectual faith in the saving grace of Jesus Christ. Swedenborg's struggles with the Godhead came to be exacerbated by his scientific endeavors, especially after he started engaging with the proto-vitalist work of the great seventeenth-century microscopists Antonie van Leeuwenhoek and Jan Swammerdam, using their research to explore the fuzzy parameters of life in the human body. Where did the soul, the *anima*, reside in the mortal flesh? Could its existence be definitively proven with a Baconian *novum organum*, with the reason, geometry, and experience that had uncovered the secrets of rocks and minerals that Swedenborg sought to discover in the *Principia?*

Here, Swedenborg's ultimate answer to the age-old dilemma comes close to Whitehead's own: in articulating a theology that posits we must first grasp, self-reflexively, a sense of our life and our place in the world as an image, as a totalizing kind of representation, Swedenborg could be said to key an aesthetic register over and above his Enlightenment epistemology and rationalist ethics. This was the heart of Swedenborg's nine-volume allegorical interpretation of Genesis: that rather than literally or historically accounting for anything, the dense symbolism in the Hebrew text powerfully pictured, or in his words, "corresponded" to, the spiritual development of consciousness as it became aware of itself and of "life as an image," as Lars Bergquist puts it, and of the ameliorative flow of divine love into (human) nature.[12] The dynamic exchange between this divine love of Creator for the Creation was a perpetual "infinite and eternal process of creation," where "an infinite variety of events may happen," a making of the foundation between heaven and earth.[13] In spite of being couched in eighteenth-century physicotheology, Swedenborg's account of this process comes very close to the metaphysics that conclude Whitehead's own *Process and Reality*. As Whitehead puts it: "[T]he kingdom of heaven is with us today. The action . . . is the love of God for the world. It is the particular providence for particular occasions. What is done in the world is transformed into a reality in heaven, and the reality in heaven passes back into the world. By reason of this reciprocal relation, the love in the world passes into the love of heaven, and floods back again into the world."[14]

To fully understand how and why Swedenborg turned to an immanental cosmology that placed divine love (and wisdom) pulsing like a sun at its center, and his further implication that representation, or the aesthetic, is not just "art" but a constitutive fact of all creation and veritably hardwired into perception, we must now turn to the early years of Swedenborg's formation, when Sweden was roiled both by the emotional fervor of Continental Pietism and, in the universities at least, by the controversial ideas of René Descartes.

Crystals, Spirals, and Stars

Emanuel Swedenborg was born in 1688, the son of Jesper Swedberg, a prominent bishop and professor of theology (the entire family would change their name from Swedberg to Swedenborg in 1719, following their ennoblement by Queen Ulrika Eleanora). Swedberg cast a long shadow over his son, as most successful parents tend to do. In the later pages of Swedenborg's journal of dreams that document his great spiritual and intellectual crisis, which began in earnest in 1742, seven years after his father's death, some of the most vivid dreams betray a longing for his father's approval of the physicotheological projects Swedenborg was then undertaking to find the soul in the human body.

Many years earlier, Swedberg had taken a theology professorship at the University of Uppsala where, starting at the age of twelve, Emanuel matriculated as a student (then a not abnormal age to enter university schooling for someone of Swedenborg's class and social standing). Uppsala was still feeling the aftershocks of the influence of Descartes, who had been brought to Sweden by Queen Christina some fifty years earlier. A subsequent rift had opened up in the faculties between, on the one hand, the older Aristotelians with their forms of Renaissance scholasticism, and the new camp of the Cartesians on the other, who championed empiricism and an epistemological premise that thinking began with an intuition of the self: *cogito ergo sum*. Swedberg the father favored the former, the son the latter, and the young Emanuel quickly emerged as an important thinker in the nascent natural sciences in early eighteenth-century Sweden.

Uppsala and Stockholm were on the fringes of a then-emergent Enlightenment cosmopolitanism. As Swedenborg began making a name for himself—working with the inventor and mechanical genius Christopher Polhelm to publish Sweden's first scientific journal, the *Dadaelus Hyperborealis,* and offering the first modern geological description of Scandinavia in his *Height of the Waters* (1719)—he found it necessary to travel to London and the Continent to meet with leading mathematicians, engineers, and astronomers, including Edmund Halley and John Flamsteed (Swedenborg even worked for a period as Flamsteed's assistant at the observatory at Greenwich, in the famous Octagon Room). His first European trip, from 1710 to 1713, marked the beginning of a lifelong dynamic of necessary travel to Europe, both for access to cutting-edge research in mineralogy and physiology, and, later, the greater freedom of the press in Amsterdam and London for his increasingly heterodox theology, which would ultimately come under censure and charges of blasphemy from the Lutheran church in Sweden.

Consistently across Swedenborg's natural science publications of this period, he evinces a thoroughly mechanistic worldview, set in motion by Aristotelian first causes. Lars Bergquist writes how Swedenborg's mechanism of this period was shaped by two fundamental assumptions: firstly, the sense that movement formed the basis of all matter, and, secondly, that this movement-impelling matter took the spiral, or the whorl, as an elementary shape—a notion that Swedenborg drew on Descartes and others to refine.[15] As author of one of the first books on algebra to appear in Swedish, Swedenborg saw mathematics—especially geometry—as particularly well suited for providing the universal means of understanding nature. David Dunér usefully observes how Swedenborg's "semiotic reading of nature" meant that "the world has a Euclidean grammar."[16]

There is strong continuity between the universalizing pretensions of mathematics to operate as a kind of abstract language that can descriptively encompass all aspects of reality, and Swedenborg's later turn to a hieroglyphic "science of correspondences" that turned all of nature into spiritual semiotics, and further hovers as a kind of *Ursprache* behind Swedenborg's history of the development of language and human communication. And yet, Swedenborg's later correspondences, as the next chapter examines in more detail, carried residues of apophatic mystical experience that caused hesitation and doubt about any symbolic system's capacity to represent the fullness of experience. It could be even said, as Jonsson claims, that it was precisely Swedenborg's intimate wrestling with and through language—the private languages of the self, its dreams, and the holy words of Scripture—that led to his existential crisis, and subsequent mystical turn.[17] Swedenborg's struggle was ultimately with the limits of representation itself.

Swedenborg's scientific writings from the period leading up to, and culminating in, his *Principia; or, The First Principles of Natural Things* (1734) are extensive and varied, ranging from crystallography, to anatomy, to both experimental and applied physics. What unites all this natural philosophy consistently is the mechanistic causality between micro- and macrocosm: the same universal laws apply to both, and one can infer perceptual truth about the one (the wave-motion of oceans) from the other (ripples in a puddle or pond). Nature was "nothing but a composite analogy," as Swedenborg puts it in one of the early issues of the *Daedalus Hyperborealis* journal, and "although one thing may appear to be different from another, it is nothing other than a composition of analogies, which is evident from the harmonic proportion, which is a geometrical and arithmetic counter-likeness together."[18] Geometry, arithmetic, and algebra provided Swedenborg with the code for descrying the analogies of nature. Dunér calls Swedenborg's syncretic mathematical methods here—of productively mixing his metaphors, of moving one analogous register into another, in order to infer a new meaning—a kind of "iatrogeometry," from the Greek *iatro-*, "combining form," and the established tradition of iatromechanism.[19]

The spatial and dimensional prospects unfolded by Swedenborg's iatrogeometry meant that he soon had to account for not only matter but also the movement within the mechanism of nature. Adapting Descartes's still-influential modifications of Galenic "animal spirits," Swedenborg proposed that all sensate matter (and not just the whirring machine of the human) vibrated, and was under constant oscillations, or "tremulations," as he put it. Using geometry, Swedenborg outmechanized Descartes: even the conduits of nerves, the pathways of thought in the brain, could be spatialized and understood through their vibrating oscillations and tremulatory patterns. Thus "it often happens that a person joins another's thought, so that he gets a presentiment of what the other does and thinks, i.e., that his membrane quivers because of the other's quivering, as one string of another's, because they are tuned to the same pitch."[20] These oscillations of thought were not unique to the human subject. Whereas Descartes viewed animals as unthinking automata, "even the thoughts of the unreasoning animals are tremulations," Swedenborg claimed.[21] The universe constantly trembled under (good) vibrations. As this movement was theorized to operate along an indiscernible plane of vital spirits or subtle matter, it let more supernatural causation come creeping through the backdoor of its own mechanization. As Hjalmar Fors observes, such "theories about vital spirits and subtle matters could blend seamlessly into mechanical philosophy but came at a price. They left open the theoretical possibility that a host of angels, devils, ghosts, and trolls inhabited the mechanical universe. And furthermore, that such entities were perceptible through the sense organ of the human brain."[22]

This vibratory movement extended to rocks and minerals, and Swedenborg's early speculative crystallography took the medieval vitalism of alchemy, with its tradition of a "living nature," and passed it through his mechanistic alembic that was derived from Descartes. Crystals, minerals, and ore sprouted and grew like foliage under the earth, not due to any occult principles of transmutation, but according to the predictable laws of geometrical space. Swedenborg's classification of minerals and geometrical crystals—he was particularly interested in how the cubes and hexagonal pyramids of pyrite grew—were influential in the nascent typologies of the earth sciences in Scandinavia, and his ideas were subsequently developed by both Linnaeus and the German crystallographer Johann Friedrich Henckel (whose own work was mutually important for Swedenborg). While Swedenborg cannot be solely credited with the invention of crystallography as a field—an exaggerated claim made by the later Romantic chemist Jean-Baptiste Dumas—his work made an important contribution, nevertheless, to a new conceptual paradigm of how the natural forms of minerals organically grew in geometric shapes, through the repetition of simple steps and the addition of forms in analogous patterns.[23]

The processes of crystallization within Swedenborg's mineralogy exhibit characteristic trademarks of Descartes's mechanistic philosophy. The junction at which they later informed his reflections on the human form, however, show the rich potential for conceptual metaphors to transfer from one domain into another, and how a corpuscular theory of materiality, steeped in John Locke, Gottfried Wilhelm von Leibnitz, and Benedict de Spinoza, could yet harbor an incipient organicism of form. The arc of Swedenborg's scientific accomplishments—from crystals and mines, into the epigenesis of the human body—anticipates the paradigm shift that Donna J. Haraway has so artfully traced in the later nineteenth and early twentieth centuries, as experimental biologists and chemists attempted to move beyond the static vitalist-versus-mechanism binary they had inherited from Romanticism, and to find a new "third" way in biological organicism (a concept that was first coined in 1928). Swedenborg's crystallography was bullular; that is, it saw matter as essentially composed of round particles, like peas or cannonballs, two metaphors that his mentor Christopher Polhelm was quite fond of, which arranged themselves into geometric crystalline figures. As Haraway writes in regards to how this conceptual imagery then gets transferred, from the mineralogical into the biological: this "globular image" that was so "important for the cell theory was an alternate early analogy for the fundamental structure of matter. These strongly visualizable forms are more than props for the imagination; they have been intrinsic to explanations of basic properties of life."[24]

Swedenborg's bullular theory was first published in 1721 in his *Prodromus Principiorum* (usually referred to in English translations as Swedenborg's work on "Chemistry").[25] The acme of this mechanistic vision, and his most comprehensive statements on cosmology and mineralogy, was the three-volume *Opera Philosophica et Mineralia*, of which the first volume formed the aforementioned *Principia*, Swedenborg's broad and sweeping attempt to account for "the first principles of natural things." Before discussing the *Principia*, if we anticipate his turn toward anatomy and physiology ten years later, and the ensuing spiritual crisis and subsequent mystical turn, it may be worth briefly considering some cultural and biographical contexts that might help account for such later deviating swerves—why, say, Swedenborg didn't stay happily ensconced in the mechanical materialism and natural philosophy of his day as so many of his contemporaries clearly did. This book will not attempt any kind of causal rationalization of Swedenborg's later visionary claims, reading his mysticism through the lens of medical pathology or the like.[26] Modern academic attempts to reduce Swedenborg's literary complexity to categorical brackets of epilepsy or schizophrenia echo the nineteenth-century treatments of "poor, mad" William Blake (as William Wordsworth and Samuel Taylor Coleridge both opined), and say more, perhaps, in a Foucauldian sense about the academic apparatus of policing the social production of knowledge than they do about Swedenborg himself. There are nevertheless some telling fissures—one could again evoke Michel Foucault here, considering the archeological discontinuities that Swedenborg presents to the Enlightenment episteme of modernity, secularity, and "rational" progress—that are already quite apparent in the first half of Swedenborg's life: discontinuous cracks, if you will, that later teem with growth in the profuse garden of his theology.[27] These cracks include his immersion in the emotional fervor of Pietism; his practice of a form of breathing meditation to induce altered states of consciousness; and, finally, his glancing, apparent interest in the shamanism found in Lapland, in northern Sweden.

Though Swedenborg's father Jesper Swedberg wrote a long, garrulous spiritual autobiography, packed with incidents and details, Swedenborg himself left behind very little of public record about his early childhood. When asked by a later religious adherent for details about his life, he responded in a brief letter that constitutes his only public narrative about himself (the extensive, private, turmoiled spaces of his dream diary and journals of spiritual experience are wholly another matter, where memory and the past are recomposed into Dantean configurations of heaven and hell). "From my fourth to my tenth year," Swedenborg wrote, "I was constantly in thoughts concerning God, salvation, and the *spiritual sufferings* of men, and several times revealed that at which my mother and father wondered, saying that angels must be speaking through me. From my sixth to my seventh years, my delight was to discourse with clergymen concerning Faith—that the life thereof is love, and that the love which gives life is the love to one's neighbor."[28]

This emphasis on suffering, on love, and even the attention to the voices of angels, betray the strong imprint of Continental Pietism in the Swedberg household. Though the state Lutheran Church of Sweden tended to regard Pietism as suspect, if not heretically dangerous to the hierarchical structure (and political power) of the church, Swedberg reserved a special place for Pietist authors in his home, in ways that were highly atypical for Sweden at that time. Swedberg would read excerpts of Johann Christian Arndt's massive *True*

Christianity (*Wahre Christentum*) at the daily evening prayers, a proto-Pietist miscellany that focused on the mystical union of Christ with his believers. Many decades later, Swedenborg's last major theological treatise would bear a strikingly similar title, the *Vera Christiana Religio* (1771) in Latin, or *True Christian Religion*. Another popular book in the Swedberg home was Christian Scriver's *Seelenschatz* (Treasure of the soul, published between 1675 and 1692). This German Pietist text emphasized how an invisible world of angels actively engaged in our daily affairs. According to Swedberg, the *Seelenschatz* was "indispensible," the most important text after the Bible: it was also devotionally read in the household of Carl von Linne (Linnaeus).[29] Because such Pietist literature encouraged a union between the self and the Divine that was wholly private, and thus in a certain regard anti-ecclesiastical, it fostered a textual tradition of typological commentary that tended to regard nature as a second book of Revelation, where God's constant love was to be ecstatically experienced by the individual. When Emanuel Swedenborg and his siblings heard and read their father's selections from Arndt and Scriver, they would have encountered older strands of unorthodox thought that hearkened back to Raimond von Sabunde and Jakob Böhme, who both believed the revelations of the Book of Nature would usher in "a new era of religious contemplation," as Ernst Benz puts it. Benz writes further how this context probably shaped a hermeneutical model for the young Emanuel in which "the character of a 'private Christianity' gave priority to the Book of Nature for the time being."[30]

Like many of his Continental Pietist counterparts, Swedenborg's father believed that nature was the external manifestation of a supernatural struggle between heaven and hell. The creaking and groaning of a household's wood furniture could be signs of divine approbation or displeasure, he writes in his autobiography; the disastrous fires that destroyed Swedberg's homes on three separate occasions were forms of persecution and revenge by the Devil, who was angry with Swedberg for converting and saving so many Christian souls. Swedberg thus raised his children in an environment where angels and spirits were always on the watch, observing each action, and the miraculous import of dreams and visions was accepted. The children were encouraged to cultivate habits of introspection and self-reflection—praying, meditating, keeping journals—where they could examine the state of their unregenerate souls and locate the signs of divine grace and salvation that lurked beneath the everyday. Ever since having a near-death experience as a boy, when his unconscious body had been dredged up out of a millpond, Swedberg professed a belief in personal guardian angels that watched and guided his every step. Later experiences detailed in his autobiography record seeing indescribable visions and hearing disembodied angelic voices.[31]

In being steeped in this kind of Pietist, visionary fervor at the same time that his proximity to the Cartesian adherents at Uppsala firmly implanted a mechanico-physical orientation, Swedenborg shares in the same pressure-cooker milieu that ultimately produced so much German Romantic thought at the end of the eighteenth and early nineteenth centuries. According to Isaiah Berlin, that later generation of Immanuel Kant, Johann Gottlieb Fichte, Johann Gottfried von Herder, and Friedrich Wilhelm Joseph von Schelling were all formatively steeped in a shared current of German Pietism. Pietism, Berlin argues, imparted an intense, interiorizing spirituality that came to be undermined by Enlightenment science, especially by the Rationalist tradition upheld by Christian Wolff (a thinker of great importance for Swedenborg). Pietism kept alive, underground, a kind of counter "mystic vitalism."[32] In regards to Swedenborg as a precursor of these German Idealist tendencies that Berlin diagnoses, it is perhaps best to consider Pietism not as a set of theological dogmas that Swedenborg necessarily ascribed to, but rather as an embodied habitus of reading and writing about the self, mediated by a scrim of sacred Scripture and the holy Book of Nature.

One quite un-Pietistical practice that Swedenborg seems to have developed at an early age was a discipline of meditative breathing that involved hypoventilation, a slowing-down of the breath to the point of cessation. His later journals of spiritual experiences detail how he would deliberately alter his breathing patterns to affect his thinking—he records how "when in childhood I wished to hold my breath purposely . . . I wished the times of respiration to agree with those of the heart, and so observed that then the understanding began to vanish as it were; then afterwards, when I wrote in imagination, that I had observed that I held my respiration as if tacit."[33] This tacit "writing in imagination," connected to breathing patterns and a sense of liberation from his understanding, became the experiential platform for Swedenborg's later visionary output. "A person in ecstasy holds the whole breath," as he records in his dream diary, "And the thoughts then being as it were absent."[34] There were no contemplative spiritual exercises in the Protestantism of Swedenborg's time that afforded anything proximate to this careful attention to body and breath, and how it modulated patterns of thought. Indeed, by and large in mystical traditions in the West—with some key exceptions in Catholicism and early medieval Christianity—it could be said that the body has played a diminished, subordinated role, particularly within Protestantism. Some scholars of Swedenborg have taken due note of his occasional references to the West Indies and contemporary spiritual traditions in Mongolia (what Swedenborg called "Greater Tartary") as evidence of suggesting, at the very least, an implicit parallel with the *pranayama* that is so pronounced in Buddhism and other Vedantic traditions.[35]

That may well be; Swedenborg does definitively go beyond the pale of Euro-Christianity in his journal when he parallels his own ongoing experiences of mystical ecstasy to the trance state of Sámi shamans from Lapland, from the far north of Sweden—he notes how such shamans manifested an "ecstasy energumen," an ecstasy of the spirit-possessed.[36] Swedenborg was likely never in Lapland himself, though he would have learned much about the Sámi from his cousin-in-law Linnaeus, who famously lived with Sámi peoples during his travels through Lapland in 1732. As a boy growing up in Uppsala, Swedenborg had also lived next door to the famous Rudbeck family of professors, where Olaf Rudbeck the Younger had amassed a significant ethnographic collection of Sámi materials that much impressed his student Linnaeus in the 1720s. Apart from the expanding colonial spaces of empire, there were few other places within Europe where emergent Enlightenment regimes of knowledge and classification could directly brush against a living indigenous religion and culture that made available models of consciousness other than, and alien to, the reigning Lockean and Rationalist paradigms (though the Gaelic cultures in Ireland, Scotland, Wales, and Brittany certainly present similar kinds of issues). In the later pages of his visionary spiritual diaries, Swedenborg records watching processions of Sámi spirits wearing traditional clothing, suggesting how the culture actively populated his imagination; he also wrote, earlier, a short Latin poem about the Sámi who "have slender limbs but wild manners."[37] It is also perhaps indicative that in reminiscences made by his contemporaries, Swedenborg was recalled to have frequently worn a "lappmudd" during the winter as he wrote: a traditional Sámi garment made of reindeer skin, worn with the skin turned outwards.[38] "The spiritual clothes itself with the natural," as Swedenborg wrote in a later aphorism about influx, "just as a man clothes himself with a garment."[39] In his earlier poetic juvenilia, the Sámi poem closes, curiously, with a description of the Sámi using animal skins for clothing so that they appear like animals;

> And they wrap their limbs in hides that they have torn from the animals;
> That which was earlier a bear becomes a garment, a covering.
> So simply is the Lapp dressed. If he were but to get down with his head bent,
> he would then be able to deceive the wild animals with a skin that is similar
> to theirs.[40]

This animal-human contiguity—a bear "skin that is similar to theirs"—anticipates John Muir's later use of Swedenborgian theology as a springboard into a reflection on the other-than-human in his famous dog-story, "Stickeen" (as chapter 4 will take up). The lappmudd-wearing Swedenborg in his summerhouse on Södermalm may not be far from Muir's dog, "a herald of a new gospel, a Swedenborg."[41]

In his 1849 biography of Swedenborg, James John Garth Wilkinson makes the interesting observation that the figure of the gyrating spiral, which comes to full flower in the nebular hypothesis of the *Principia*, echoes pre-Christian thematics in ancient Norse sagas about the cycles of the Yggdrasil tree.[42] Even though no volumes of Scandinavian mythology or folklore appear in Swedenborg's extant library, he would have been familiar with the basic shape of the Icelandic eddas from the Rudbecks and other *polyhistors* in Uppsala. Wilkinson's offhand remark suggests how Descartian spirals and vortices may have absorbed other valences when situated on the periphery of the Enlightenment, in the hyperborean margins of Sweden's emergent scientific cultures.

Ultimately, however, one doesn't necessarily have to turn to the more exotic spaces of Lapland (or farther afield to Asia, as Marsha Keith Schuchard has argued) for locating alternative models of consciousness that would have been available to Swedenborg.[43] As Friedemann Stengel cogently points out, there is plenty of exoteric content in the so-called Rationalist tradition that would have offered Swedenborg any number of alternative epistemologies that many scholars have misattributed to more esoteric sources, such as the Jewish Kabbalah. In terms of the oft-cited passage above where Swedenborg observes the connection between breathing and thinking, and a "writing in the imagination," it is interesting to note some of his earlier notations made in his expansive commonplace book from the 1740s, the so-called "Philosopher's Notebook" manuscript. Filled with hundreds of quotations culled from a variety of sources, then (re)arranged under conceptual headings and indexes of his own devising, the "IMAGINATION, MEMORY" section features a lengthy excerpt from Christian Wolff's *Rational Psychology* (1734), a text that, tellingly, features one of the first uses of "psychology" in its modern scientific derivation.[44] Swedenborg takes note: "for exercises whereby the force of the imagination is intensified, we must first choose the night-time or evening; and especially that time when there are few objects to strike the sensory organs; or the eyes must be closed so that no light may enter."[45] The imagination was to be exercised in the liminal spaces of twilight; the closing of the eyes was, literally, in the Greek, a Μυεσις, the basis of the English word "mystic"—which, as

Ralph Waldo Emerson points out in his later essay on Swedenborg, was but "a getting out of the body to think."[46]

Whatever its sources, Swedenborg's disciplines of meditative (pranayamic) breathing, of Sámi *exumen energem*, and the fervor of a particular kind of Pietism can all be seen as attempts to get at "the paradoxical existence of a 'something' which simultaneously cannot be sensed (from the point of view of the empirical exercise) and can only be sensed (from the point of view of the transcendent exercise)," in Gilles Deleuze's words. In *Difference and Repetition*, Deleuze outlines three ways that a body might grasp, or experience, the ultimate ground of sensation that provides the field of differentiation which makes all meaning possible. One is through psychotropic drugs that displace the self; another is through the bodily experience of vertigo, "a depth in itself or that intensity in itself at the original moment at which it is neither qualified nor extended." And yet a third way is through the "sensory distortion" opened up by "a pedagogy of the senses, which forms an integral part of 'transcendentalism.'"[47] Parallel and coterminous with Swedenborg's mechanical mineralogy, he seems to have actively practiced such a transcendentalist "pedagogy of the senses." It may come as no surprise, then, that his later turn to the physical body in his studies would sharpen an intensity of the sensible substrate, a feeling that "in a word, absolutely everything in nature, from the smallest to the greatest, is a correspondence."[48]

No "pedagogy of the senses" is on display in the *Principia* other than a fully Lockean empiricism, as the text unrolls an elaborate theory for the creation of the universe and subsequent formation of geological layers in the earth. The speculative material in the *Principia* was meant to be read as a counterpoint to Isaac Newton's own *Principia Mathematica* (Swedenborg, like Blake and other Romantics, particularly differed with Newton on the question of the vacuum), even as the more practical material pertaining to smelting and ore extraction became quickly adapted, translated, and used by inventor-entrepreneurs in burgeoning industrial economies, including (later on) James Watt in England.[49] Building on Descartes's work on spirals and the vortex as ur-forms of matter and motion, the *Principia* articulated what has been subsequently called by historians of science the "nebular hypothesis," a speculative cosmology accounting for the origins of the solar system and universe. In attempting to justify a conviction that the beginnings of the universe was a state of energetic, unformed chaos "common to the sun and the planets," Swedenborg turns to ancient classical poetry, quoting at length from Aristophanes and Ovid to argue that such voices evince a common truth that "universal reason" had once been capable of perceiving. Swedenborg uses Ovid's chaos to describe this primordial mass of energy:

Once was the face of nature, if a face
Rather a rude and undigested mass
A lifeless lump, unfashioned and unframed
Of jarring seeds; and justly Chaos named.[50]

Swedenborg's extensive turn to classical poetry at this juncture in the *Principia* is striking in a work almost eight hundred pages in length (in its English translation) that is otherwise much preoccupied with geometrical formulas, tables, and precise proofs of Swedenborg's various experiments with magnetism. It illustrates and typifies, for one, Alfred North Whitehead's observation about the birth of modern science in the post-Renaissance period: namely, that it inherited from the revival of Greek traditions an anti-Rationalist bias that insisted on viewing the cosmos in dramatic (or poetic) terms. The unfolding of nature's secrets proceeded within recognizable narrative and rhetorical expectations; observations of nature's phenomena were shaped by precepts of plot and characterization.[51]

Swedenborg took the ancient view of the cosmos as a giant egg and refined it through his materialistic, Cartesian view of causal motion and spiraling matter. The revision of the universe-as-egg further signals Swedenborg's great indebtedness to Thomas Burnett's *Telluris theoria sacra* (The sacred theory of the Earth), first published in 1681, a sally that was meant to square the new Baconian science with biblical revelation, particularly when it came to the account of the Flood in Genesis. Swedenborg had begun reading Burnett as early as 1710, and he would refer back to the Cambridge Neoplatonist long after the *Principia* with his *Worship and Love of God* (1743), an experimental prose-poem that marks his transition from natural science into visionary theology (which will be more fully addressed in the final section of this chapter). Swedenborg reformulates Burnett's organicist cosmos-as-an-egg into an image of planetary birth, with the planets streaming out of the mass of their "father," the sun. This marks a crucial metaphorical distinction from Descartes's vortical theories (otherwise so omnipresent in Swedenborg, alongside Leibnitz's monads), as Descartes had viewed the planets as extinguished, dead stars that had been sucked into a vortex caused by the sun. Burnett, and then Swedenborg following him, envisioned rather a vital, living continuity between planets and the sun. Also modifying Swedenborg's cosmology were the magnetic theories of William Gilbert—volume two of the *Principia* substantially cites Gilbert's work from the prior century. Gilbert saw the seemingly universal characteristics of magnetic attraction, polarity, and orbiting as strange, wonderful signs that "the whole world, the globes, all the stars are alive, animate," as David Dunér describes Gilbert's position, one that clearly adumbrates the "vibrant matter" of the new materialists: "Why should not the stars have souls, if worms, ants, roaches, plants and morels have? . . . [T]he magnet is alive, as a part of the living mother earth and her beloved offspring."[52]

The birth of the planets—the heart of Swedenborg's nebular hypothesis—is graphically portrayed in a series of stunning figures that accompany the text (see fig. 1). These images constitute one of the first scientific attempts to describe the solar formation whereby planets are created out of the whirling, circling chaos of a nebular mass. The plates were not simply illustrative or ancillary to Swedenborg's text; he was aware of the strange newness of his

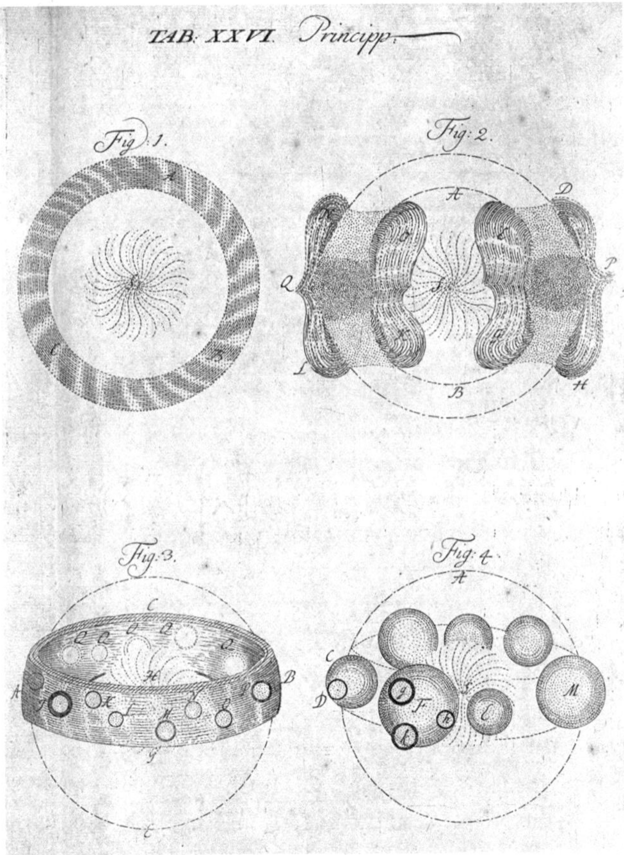

FIGURE 1. Table 26 from *Principia rerum natralium,* by Emanuel Swedenborg. (Dresden and Leipzig: F. Hekel, 1734; photograph courtesy Yohana Junker, 2016)

claims and the dilemma of trying to depict something like deep time. He uncharacteristically explains to the reader how "a thousand words will not convey the idea which may be derived from a single representation . . . so that we shall now have recourse to figures to delineate this chaos or solar crust, and its state of separation."[53] The attempt to visualize the chaos of creation through word and image builds on the tradition of showing the "hyle," or primal substance, that first appears in medieval manuscripts as an abstract plane (sometimes charmingly held dangling from the hand of God). Swedenborg would have also known the abstract figures appearing in the Cambridge Neoplatonist Robert Fludd's *Natural History of Both Worlds, the Macrocosm and Microcosm* (originally published in Latin as *Utriusque cosmi maioris scilicet et minoris: Metaphysica, physica, atque technica historia*, 1617).[54] In Fludd's seminal repository of early modern alchemical and esoteric lore, the space before the creation of the world is visually rendered as a floating square of smoky black cross-hatchings: looking oddly like a canny anticipation of Kasimir Malevich's later Supremacist paintings, as Raphael Rosenberg points out.[55]

As Swedenborg goes through a step-by-step account of planetary formation and gravitating vortexes and gyres, beneath the figures and images he again turns to Ovid and Aristophanes by repeating their words on chaos and the creation of the night. By contrast, "Mosaic Philosophy," as Swedenborg denotes the entire Judeo-Christian tradition, gets but a brief passing mention, and then only to note how the "chaos of Moses" described in the first Book of Genesis rightfully accords with the older Abyss and Tartarus described by the more ancient Greeks. The displacement of a literal interpretation of Scripture had already begun, nudged by this neoclassical science.

Swedenborg attempts to balance the poetic descriptions with a number of contemporary and modern astronomical observations of stars that seem to have either suddenly appeared out of nowhere or wavered and then permanently vanished: thus showing, Swedenborg claims, "the origin of the planets actually imaged forth to the eye" in the birth and death of various stars. Here, Swedenborg was not just touching on a revolution in cosmology, but broaching a radical new understanding of time, of the implied ephemerality of our own solar system. This moment of ongoing solar creation in the *Principia* is quantitatively far from the fixed Aristotelian Great Chain of Being, which placed the Earth firmly at the center of the universe, and stands somewhere between the upset of the earlier Copernican and Galilean revolutions and later attempts to comprehend the complexity of cosmological deep time, and the further prospects of a multiverse.[56]

This was certainly not how the *Principia* was received in its day, however. Anders Celsius and others in the scientific vanguard at the Swedish Royal Academy scoffed at Swedenborg as "an old-fashioned Cartesian and speculative particle philosopher."[57] More recent historiographies of science have tended to agree; Sten Lindroth notes how Swedenborg's "cosmological theory was interwoven with wild speculation and partly characterized by poetic extravagances," even as it critically foreshadowed the definitively modern nebular hypotheses of Kant and Pierre-Simon Laplace.[58] But it was precisely the "poetic extravagances" of the *Principia*—its turn toward classical poetry, the rhetorical reliance on the imagery of its plates—that caused a later Romantic author like Edgar Allan Poe to gravitate toward the text; such extravagences offered "soft" places that one could press on, pushing to reveal how Swedenborg was able to transform a premodern entelechy (the use of Gilbert's animated magnets cited earlier, for example) within the Cartesian machine of man-and-nature. As such, Swedenborg's "eggy" natal birth of planets is a kind of back-to-the-future disruption of scientific positivism, in the way that Bruno Latour has envisioned a future posthumanist nature going "back" to some of the premodern notions of micro-macrocosmic harmony that characterized so much science of the sixteenth century.[59] Swedenborg goes even further back, of course, in the *Principia* to Ovidean chaos, which amplifies ideas that had first appeared in Plato's "Timaeus," a text that became a "blueprint for 2,400 years of subsequent Western cosmology," as Mary-Jane Rubenstein observes.[60] Whitehead also notes that although there is much "foolishness" in the "Timaeus," its incipient organic doctrine of forms and argument for how creation perpetually comes into being makes the implicit physics of Plato's text "two thousand years ahead of its time," in stark contrast to the Newtonian creation ex nihilo that afforded "no provision for the evolution of matter."[61] Celsius and others who ridiculed the backwardness and baroque extravagance of the *Principia* were still firmly embedded within a Newtonian mechanistic framework that Swedenborg was already starting to transform, in part through this attention to an implicit organicism in the Neoplatonic tradition. Tellingly, over the next decade, starting in 1739, he fills his commonplace *Philosopher's Notebook* with many quotes and references from the "Timaeus," including how "in this universality of nature, there is something which is permanent and intelligible, and which is the exemplar of those things that are generated and *are engaged in a perpetual flux of mutations.*"[62]

Such flux and mutation continued in the ongoing creation of minerals and rocks underground, long after the originary explosion of the solar mass.

The *Principia* often expresses a "feeling of wonder" at the sheer profligacy and growing energy of metals, crystals, and ores in the earth: something that Swedenborg did not simply contemplate in the abstract, but intimately knew from his years of crawling and climbing through various mines in Sweden and northern Europe as part of his duties as a royally appointed assessor on the Swedish Board of Mines. As Bergquist puts it, Swedenborg was coming to see the mineral kingdom as but an extension of the same life force that governed the plants and animals he was simultaneously studying above ground: "[M]inerals and philosophy were connected with each other . . . the principles governing the vegetable kingdom applied also to the realm of minerals."[63] The *Principia* praises the "rich veins of minerals that spring and grow," and metaphorizes underground spaces as "a general mother that yields and renovates the seeds of the vegetable kingdom, and as a nurse supplies all kinds of plants and trees with their peculiar juice."[64] The key words for Swedenborg's descriptions of minerals and metals at this point are "renovating," "metamorphosis," and "revolution," and one can locate here in the *Principia* the ancient belief in minerals as living, growing tissues feeding into a (pre)Romantic predilection for organic forms and process.

Ghosts in the Mine

Swedenborg's scientific output up to this point was shaped by the decidedly practical bent of Sweden's early modern scientific culture, based on networks and flows of knowledge distribution between the emergence of natural philosophy at the university—especially in Uppsala—and the government-appointed Board of Mines. Made an assessor extraordinaire to that body in 1716 by currying favor with the irascible King Charles XII, Swedenborg would end up as a civil servant of the powerful board for thirty-one years. The inventions and mechanical plans that populate the pages of the *Opera Philosophica et Mineralia* were designed to improve the conditions of mining operations all over Sweden, but especially at Falun, in the province of Darlana to the northwest of Stockholm—then the largest source of copper for all of Europe, and a key economic engine for the national economy. The open-pit mine at Falun was already in the eighteenth century a vast environmental catastrophe, and it formed an important backdrop for two of Sweden's emergent natural scientists: Swedenborg, and slightly later, Linnaeus, who married Swedenborg's first cousin, Sara Moraeus, the daughter of the chief doctor at Falun. The family fortune—both Swedenborg's and that of Linnaeus's in-laws—was closely tied to stock investments at Falun and smaller mines elsewhere. When the Swedberg family became ennobled in 1719 and changed their surname to Swedenborg, the new patronym was derived from *Sveden,* the ancestral farm outside of Falun (where, coincidentally, Linnaeus married Sara Moraeus in 1739). On the new Swedenborg family heraldic coat-of-arms, what looks like an erupting volcano in the lower right corner is meant to indicate a metal foundry and its smoking chimney, or a burning mountain—symbols of the mines and interior powers of the earth that fueled the Swedenborg family finances.

In spite of the wealth it generated, the extensive toxic byproducts from Falun's enormous mine came to poison the surrounding landscape. The high sulfur content of copper ore, accompanied various heavy metals (such as lead), have left an indelible mark on the otherwise bucolic Dalarna region (still the heart of Sweden's traditional folk culture, and a more recent site for the recrudescence of right-wing ethno-nationalism). Even today, although copper production ceased many years ago, at least eight lakes around Falun have not recovered from the chronic acidification caused by the dumping of sulfur and poisonous metals into different local waterways.[65] As Falun was constitutive to Sweden's modernizing and nationalizing projects, we might read the mine as exhibiting one of the hallmark characteristics of the Anthropocene: that of its spectrality, the uncanny ways it becomes haunted by a future of toxicity produced in its present. In Nils Bubandt's anthropology of sites of contemporary environmental catastrophe in Indonesia, he argues that such entanglements between human-caused and naturally induced disasters (like Fukashima, Bhopal, or Chernobyl) indicatively disturb the boundaries between human actants and the forces of nature. "The increasing impossibility of distinguishing human from nonhuman forces, the *anthropos* from the *geos,*" Bubandt proposes, produces a characteristic "undecidability" that haunts human imaginaries in the Anthropocene.[66] As I presently write this (in November 2018) under a haze of thick smoke, are the horrific wildfires currently raging (again) across northern California caused by climate-change-induced drought? Or is it the chronic mismanagement of federal forest lands, their underfunding, under the Trump administration? Or some combination of both? During the catastrophic Carr fire earlier in the summer that destroyed areas around Mount Shasta to the north, the conflagration inadvertently brushed against an older strata of sacred storytelling about the revengeful Loon Woman who burns various creatures, and is part of the foundational story-ing of the Shasta mountain itself for the Achomawi and other area indigenous tribes: were these new fires a kind of spiteful return of Loon Woman?

In Bubandt's reading of these sorts of convergences between the human, natural, and potentially supernatural, instead of the Anthropocene performing an ostensible extension of a secular mode of disenchantment, it has rather become inspirited with the uncanny, and the surprising return of spirits and ghosts. Bubandt means this not only in a more literal sense for specific sites like the mud-flat disaster in East Java he has extensively studied, but also in general terms of how we conceptualize our present. The Anthropocene (which is marked by conceptual beasts of strange configuration, from James Lovelock's older Gaia to Haraway's recent Cthulhu), is "haunted by spirits,"

> . . . [and] a spectrality that characterizes the Anthropocene more generally. The Anthropocene, after all, invites us to imagine a world in which an alien geologist from the future detects in the strata of the ground evidence of the presence of humans long after we have gone extinct. This science fiction–like character of the concept of the Anthropocene opens up to a retrospective reading of the current moment, a 'paleontology of the present' in which humans themselves have become geological sediments or ghosts. In the Anthropocene, life is already geologic.[67]

The mine at Falun was haunted geology, even more acutely so in Swedenborg's day when he spent hours crawling and exploring its dark recesses in his role as assessor. As Fors has convincingly shown, beliefs in trolls, spirits, and "keeper entities" of mineral and metal deposits persisted among top scientists and engineers on the Swedish Board of Mines well into the eighteenth century. Swedenborg's friend Count Gustav Bonde, who was standing president of the Board of Mines for many years (and who would later own many of Swedenborg's visionary theological writings, and seems to have been clued into the secret, early on, that Swedenborg was their anonymous author) practiced alchemy in his private chambers. The great mineralogist Axel Fredrik Cronstedt, the discoverer of nickel, while publicly a skeptic of the supernatural, recorded strange encounters with evil spirits that spooked his horses—an attempt, Cronstedt thought, of a keeper-entity in nature trying to prevent the Board of Mines from building new ironworks in a forest. Urban Hiärne, a rival chemist of Swedenborg's mentor Christopher Polhelm, while far away from his home, had a premonition that his wife was giving birth to their son after experiencing sympathetic birth pains in his back and loins. When we collate such incidences of the paranormal, rather than Swedenborg's later clairvoyance and angelic communication being seen as an anomaly—or the regressive persistence of premodern superstition—Swedenborg was perhaps more representative of the Board of Mines than one might otherwise think. The eighteenth-century board generally held "a syncretic worldview," Fors writes, "merging folk beliefs with late Cartesianism and Lutheranism."[68]

Falu Gruva, the great mine at Falun, was rumored to be watched over by a particular spirit, a beautiful *gruvfrun* (mountain-wife) who was the *bergsrået* (keeper of the mountain). When miners saw the *gruvfrun* attired in shining bright clothing, all would be well; when she was seen wearing black, however, it forebode an accident or imminent cave-in. Legend also recounted how the mining site had first been discovered by a frisky goat, Kåre, who came back to its owner with red-tipped horns, cluing him in to the presence of valuable ore. The origins of Falu Gruva are thus deeply entangled with the other-than-human, as the mine's post-industrial afterlife (of ecological devastation) marks it as part of the Anthropocene. In neither his private nor public writing, however, does Swedenborg suggest an interest or engagement with these aspects of Falun's (super)natural history. In contrast to Bonde and the other alchemically inclined scientists on the Board of Mines, Swedenborg's *Principia* dismisses alchemy as a "farrago of fictions," and warns the reader to "believe nothing but palpable experience."[69] The mine nevertheless registers itself as a spectral sort of presence in an early poem that Swedenborg composed in 1716, a text exhibiting what Jacques Derrida termed the "hauntological": a temporal, historical, and ontological disjuncture that reveals a "lost future" of modernity and its various forms of capitalism.[70]

Before turning to Swedenborg's poem about the mine, there are two further moments in Falun's history that are important to weave together in the background. These events, where human activity at the mine distinctly "cross-hatches," as Dipesh Chakrabarty would put it, the *longue durée* of geologic time and natural history, occurred in such a way that their very "undecidability" (Bubandt) of causation creates a kind of spectral indeterminacy. In 1687, the year before Swedenborg's birth, after several months of ominous rumblings (and apparitions of the lady of the mine in black), the mine collapsed on itself in a spectacular fashion, causing a vast pit to open up, mawlike, in the landscape (which is how Falu Gruva still appears today). Miraculously, because it happened to be the annual midsummer holiday, most of the hundreds of mine workers were not on-site, but down in the town of Falun celebrating and carousing—one of only two days out of the year (the other being Christmas) when the mine would have not been worked around-the-clock. The cave-in appeared to be the fault of overmining (tunnels had been dug there consecutively since the tenth century), although many were quick to interpret it as either a merciful judgment from God (because so few had died) or the activity of the lady of mine who had been angered by the overreaching greed of the humans. Something truly uncanny occurred thirty-three years later when miners uncovered the perfectly preserved corpse of a miner who had been buried in the earlier cave-in. The toxic byproducts of copper extraction and smelting had perfectly embalmed the corpse in its antiquated clothing, preventing it from decomposing. The youthful body was subsequently put on display in a vitrine at the mine, where the miner's former fiancée—by this time a very old woman—recognized her former youthful lover, identifying the corpse as that of Mats Israelsson.

This odd warping of time, enabled by chemicals and touched by love, became a favorite tale for later German Romantics, most famously as retold by E. T. A. Hoffmann in his "Die Bergwerke zur Falun" (The mines at Falun, first published in 1819). Ecocritics such as Kate Rigby have read Hoffmann's tale as "disclosing the potential complicity of mine mysticism with the increasingly industrial exploitation of the natural world," and how there was a "growing incompatibility of the romantic philosophy of nature with the constraints of bourgeois society, in which the technological mastery of nature was conjoined with a libinal repression of the embodied subject."[71] In this regard, Israelsson's body (in Hoffmann's story, the miner is renamed Elis Fröbom) echoes the out-of-time disruptive presence of the iconic monster in Mary Shelley's *Frankenstein,* which had been published the year prior, in 1818: a novel that imagines, for the first time, the capacity for humankind to completely destroy itself.[72] The textual genesis of Shelley's *Frankenstein,* moreover, is part of a history of climate disruption, with Shelley initially drafting the story as a form of indoor entertainment during the long, dreadfully rainy summer of 1816 that had been caused by the volcanic explosion of Mount Tambora in Indonesia. The bodies of the monster in *Frankenstein* and the dead miner in Hoffmann's tale about Falun are both spectral, in Bubandt's sense, in their violation of normative time frames (especially that homogenous empty time of secular modernity), and the indeterminacy of their respective creations. Have they been formed by human folly, the forces of nature, or something supernatural? The monster of *Frankenstein* and the corpse of the "Mines at Falun" adumbrate the uncanny of our own Anthropocene.

Even though Swedenborg's poem was written three years prior to the exhumation and spectacular public display of Israelsson's body, we can find a similar tone of ambivalence in its staging of human relationships to the nonhuman matter of the mine. Written in Neolatin, the poem begins by narrating the terrifying method that Swedenborg and other workers would use to descend into the mine by holding onto a small, winched bucket that plunged, elevatorlike, hundreds of meters into the depths. Like much of Swedenborg's early writing, the poem relies upon a network of classical allusions in order to paint the recesses as a kind of Hades:

Behold, a wonderful thing! I glide down from the upper world in a bucket,
thus hanging, I am brought all the way to the dark shadows of death.
But, as I moved to and fro hanging in the middle of the air,
it was pleasant for me to sing holy hymns.
It was pleasant thus to weigh this poor and fragile life,
which all depends on the power of a rope.

The poem then proceeds to catalogue the workers in the mine below:

Behold—in the recesses of the mine the band of Hades hurriedly rushes along,
with dark faces wondering at me and my followers.
Some of them roll rocks, some sweat in the dust,
this one throws down wood, another brings torches.
Yet another enters the recesses, someone puts out from a hole
a face that is so burnt, so that even dark Charon himself would laugh.
Another with great effort climbs quivering ladders, yet another is squatting in
    a vessel;
the vessel only holds half his body.
This one is working himself through winding veins, another goes round,
for he leads small horses around in a circle.
What more shall I mention? All is full of hard work and skill.
Skill is not inferior to the material, nor is the material inferior to the skill.[73]

The poem's last two lines encase a curious double negative that equalizes the men and their work and the natural materials before them in the mine: a recursive gesture that could also be said to preoccupy the ambivalent ending of Hoffmann's "Mines at Falun." The poem is not a triumphant celebration of mechanism's ingenious mastery of the mine, as it ends on an equivocation of a "material" not inferior (i.e., below) to the human subjects who work the vein.

In Swedenborg's declarative refrain "Behold!" we might hear a distinctively theological sort of echo. In the original Neolatin, the word here is either *en mirum!* (A miracle!) or *ecce!*, (look!), the latter being the same apodeictic word used in the Latin Bible translations that Swedenborg read of the famous moment when Pontius Pilate presents the scourged body of Christ: "Ecce Homo!" (Behold the Man!).[74] As Serenella Iovino argues, "behold" evokes both the transcendent and the immanent, and often implies the all-seeing power of that ultimate "hyperobject," God. To behold can be a "mystical stance," an "invitation to reenchant the world by holding together the entangled, luminous fragility of all beings."[75] Swedenborg's Behold! in the poem draws immediate attention to the mortal fragility of his own human body, "the dark shadows of death" that then invert into pleasant hymn-singing in the swinging, plunging bucket (which seems to also be a clever metonym for the poem itself, made up of thin lines descending down the page). As a locus of transcendence (the vertical shaft, depending on the power of the rope) and immanence (the material of the mine *not* inferior to the skill of the miners, laterally in the tunnels below), this might be a site for an uncanny haunting by the future, and delineating the spectrality of the Anthropocene: all the more so when we recall Falu Gruva's central role in the modernization of Sweden. In the poem, though Swedenborg's body glides down from the upper world, he does not narrate his return from this mineralogical Hades. Perhaps the ascent and return is figuratively provided by the corpse of Mats Israelsson, exhumed three years later and placed on display in the glass case—a body as equally made by the chemical and natural forces of the mine itself as by the human error that instigated the great collapse of 1687 (there is the prospective third causal agent here, as fictionalized by Hoffmann, with the seductive lady of the mine who lures and traps the miner). Israelsson's spectacular corpse embodies what Bubandt has called the "necropolitics" of the Anthropocene, modifying the term from Achille Mbembe, its life-and-death effects of "ruination and extinction": "in the Anthropocene, necropolitics operates under the sign of metaphysical indeterminacy rather than certainty, unintended consequences rather than control."[76]

The body returns for Swedenborg, so to speak, shortly after he commenced his later *Opera Philosophica*, with its nebular hypothesis and smelting techniques. More or less immediately upon publication of the first volume of the *Opera*, Swedenborg turned his restless attention to the human body. Swedenborg's manuscripts in the subsequent decade, leading up to his visionary crisis year between 1743 and 1744, are extensive and varied, from standard essays in the genre of physicotheology (*On the Infinite*, 1735), to unpublished treatises on brain physiology, to his last two great published works of natural science, the so-called *Animal Kingdom* (1744–45) and *Economy of the Animal Kingdom* (1740–41) that labored to prove the existence of the soul in the human body.[77] It was this intimate turn to the human form—leaving behind, figuratively and literally, Falun and the mines of northern Europe for the labyrinths of blood and cranial circulations in the body—that precipitated a widening of the cracks in Swedenborg's Cartesian mechanism.

Bodies, Brains, and Eggs

As preparation for the *Economy* and the *Animal Kingdom*, Swedenborg immersed himself in the microscopic worlds opened up by the great anatomists

Marcello Malpighi and Jan Swammerdam. It is possible, too, that he observed (and maybe took part in) dissections of human bodies in Paris in 1736, and perhaps saw Pourfour du Petit conduct his famous experiments on the lesions of cortexes in dogs. Some of his unpublished manuscripts on the brain from this period—essayistic sketches for work that was planned to be part of the *Economy* and *Animal Kingdom*—contain prescient insights about the motor functions of different regions of the cerebral cortex. This brain research was "totally unprecedented and was not to appear until well into the nineteenth century," writes the historian of science Andrew Gross, as no other suggestion of somatatopic organization of cortical functions surfaces in scientific literature until the definitive experiments of Gustav Fritsch and Eduard Hitzig in 1870, some one hundred thirty years later.[78] Perhaps even more uncannily prescient, as Stanley Finger points out, Swedenborg implied that the frontal lobes of the cortex were involved in imagination, memory, and thought—standard views for neuroscience today, whereas most eighteenth-century anatomists regarded the cortex as an insensitive rind.[79] In these manuscripts, Swedenborg slipped out of the medical paradigms of his day, prehending (in Whitehead's sense of the word) the embodied, physiological ways that cognition was *in* the body.

Some of this insight seems to have been facilitated by a growing kind of "pattern recognition," whereby Swedenborg was able to broaden and amplify his field of metaphorical associations in ways that productively mixed contexts and linguistic targets, to use George Lakoff's terminology, as he moved ideas to-and-fro between natural philosophy, physiology, hermeneutics, and theology.[80] His commonplace *Philosopher's Notebook* from this period shows Swedenborg engaging readings in Plato, Aristotle (who had said, after all, that the best thing by far was "to be master of metaphor"), Augustine, Descartes, John Locke, and Nicolas de Malebranche—a roster of authors who betray, perhaps, Swedenborg's fatigue with the Rationalist tradition of Leibnitz and Wolff that had earlier buttressed all the mechanism of the *Principia* and other volumes in the *Opera*. Swedenborg's new substantial turn to Malebranche is particularly significant. The seventeenth-century French theologian had laid out a dynamic relationship between body and soul, mind and nature, in a text that appears to have colored Swedenborg's emergent ideas about correspondences and analogical thinking. Malebranche had equated the psychophysical entanglement of mind and body with the cosmic interpenetration of the Divine into created matter, which confirmed for Swedenborg that by exploring the microscopic details of the circulatory and lymphatic systems, or the structures of the brain, one could intuit larger truths about the universe as a whole.[81]

Thus, it is in an extended section on the circulation of blood in the kidneys and their relation to the nervous system that Swedenborg crystalizes for the first time his thinking about correspondences, the central idea that later generates a "language of things" in American Transcendentalism. That this idea makes its debut in Swedenborg's thinking and writing about the interstices of the body, and not within theology, biblical hermeneutics, or metaphysics, is significant. The circulation of the blood in the liver, writes Swedenborg, contains an "astonishing" symbolism "pervading the living body," and that "a spiritual or theological dogma" is concretized in every physical aspect of nature: "one would swear that the physical world was purely symbolical of the spiritual world."[82]

The centrality of the body in these final scientific epics becomes the measuring stick for Swedenborg's subsequent spiritual world, with all of the cosmos arranged into a mystical vision of the universe as a single human organism—the *Maximus Homo.* The body as a concrete site for the manifestation of spiritual truth also distinguishes Swedenborg from his earlier dependence on Leibnitz, Spinoza, Wolff, and especially Descartes: ultimately, their mechanistic frameworks for discussing the Divine in mathematical terms were too abstract and suprahuman. In a curious way, through his turn to the body as a unit of measurement in the perceptual field, Swedenborg can be seen as part of a broader revolt against seventeenth- and eighteenth-century science and its so-called "fallacy of misplaced location," in Whitehead's terms.[83] In the dream diary that Swedenborg began to keep during this transition period, the earlier mechanism morphs into a monstrous nightmare—in the first dated entry from 1744, he writes of "standing by a machine which was moved by a wheel; its spokes entangled me more and more and carried me up so that I could not escape."[84]

The body-brain thinking in the *Economy* and *Animal Kingdom* precipitated the full existential crisis that unfolds in the pages of his dream diary. In the midst of this turmoil, Swedenborg managed to publish part of one final significant work, *The Worship and Love of God,* which was written, perhaps not coincidentally, at the same time that he acquired his Södermalm property and made plans for its spacious gardens with the summerhouse (*lusthus*). A hybrid prose-poem that takes on the original Garden of Eden story, *Worship and Love of God* participates in the so-called "hexameron" literary tradition that John Milton's *Paradise Lost* most famously represents. Swedenborg's own garden theodicy reworks his earlier scientific cosmology into literary terms, transforming the nebular hypothesis into a planetary drama of birth and generation.

The central problem for Swedenborg's Adam, "the first-born," refracts Swedenborg's own dilemma: how does one perceive the spiritual within the natural, in the terrestrial Garden of Eden that surrounds him? In the didactic manner of Milton's Raphael and the other angels of *Paradise Lost,* Adam is usefully instructed by a series of "heavenly intelligences" that flash inside the theaterlike space of his brain. The Divine is critically an "influx" into nature, these intelligences report, and Adam should attempt to perceive how

> all these things [in nature] are derivative veins from that one single Fountain. Supreme things therefore, or things superior in order, inflow into inferior things, and these into ultimate things, but not vice versa. . . . By this order follies themselves are reformed into intelligences, and insanities into wisdoms; slime is changed into the brightest gem, and dust into shining gold; the innate darkness of nature becomes resplendent as light; our acts become pieties and virtues; and moreover all things succeed according to our wish and demand.[85]

The transformation of "slime into the brightest gem" suggests Swedenborg's earlier mineralogy, shaped by the mine at Falun, and its incipient organicism—the sense that minerals and crystals were dynamic, fluid, and changing. It also evokes the alchemical, a frame of reference that Swedenborg was deeply ambivalent about. Adam's spiritualized perception further anticipates the famous parabolic ending of Emerson's later "Nature" essay, where Emerson enjoins the reader to "build your own world" so that "disagreeable appearances, swine, spiders, snakes, pests, madhouses, prisons, enemies, vanish; they are temporary and shall be no more seen."[86] In both cases, Swedenborg's and Emerson's inflated hyperbole is an attempt to push through things to some spiritual source beyond.

Just as Emerson's "Nature" originally opened with a Neoplatonic quote (misattributed) to Plotinus, Swedenborg's "one single Fountain" of divine influx that recurs in the *Worship and Love of God* comes out of Neoplatonic emanation theories, those articulated by Plotinus in particular. Swedenborg read Plotinus directly, but also critically absorbed his ideas second-hand through Cambridge Platonists like Burnett in his *Sacred Theory of the Earth,* and in the work of Robert Fludd and Henry Moore.[87] In contrast to the classic Deist view of the Godhead as a detached watchmaker or mechanical inventor, observing the wheels of creation from afar—a view that thoroughly permeated his earlier science—Swedenborg's new fluidity and flow of an emanating Divine, which he retains into the subsequent theology, is best characterized as *panentheistic:* a kind of compromise struck between the flattening effects of monistic pantheism (which rendered matter synonymous with God), and the static fixity of Neoplatonic dualism (which tended to subordinate nature as the baser, inferior shadow to a higher world of ideal forms). The insertion of the Greek *en* (meaning "becoming") into pantheism, to form a pan*en*theism—a becoming of matter, an emanation of the natural out of the theistic, or spiritual—is a specific neologism that was first coined by the German Romantic theologian Karl Christian Friedrich Krause. Like many other German Idealists, Krause became deeply entangled with Swedenborg's thought.[88] Kate Rigby's more recent ecocritical resuscitation of this somewhat archaic term further suggests the utility of panentheism for situating Swedenborg within new materialist ecocriticism.[89] In both consonant and dissonant sorts of ways, Krausian/Swedenborgian panentheism resonates with Timothy Morton's attempts to overcome the substantialist monism of Romantic pantheism without resorting (back) to a Cartesian or Idealist bifurcation of mind and nature. In *Ecology after Nature,* Morton draws on Friedrich Schiller's ironic stance toward the aesthetic to argue we must hold the position of a "melancholy dualist," and that instead of embracing pantheism (with its potentially dangerous ramifications into "Blut-und-Boden" political ideologies) we need a term like ~~pantheism~~—crossed-out, but still visible—to (ironically) keep a transcendental crack open in the matter of nature, so that we are not beholden to "an authoritarian view that there is an external thing [nature] to which the subject should bow."[90] Though Morton's more recent work in object-oriented ontology seems to have abandoned this earlier negotiation with Cartesian idealism, his ~~pantheism~~ nevertheless persists as strange kin to the forms of panentheism that Krause's original work continues to generate: both panentheism and Morton's ~~pantheism~~ are united by an attempt to mitigate the extremities of (transcendental) idealism, on the one hand, and the deterministic reduction of mind to matter, on the other.

The panentheism of Swedenborg's *Worship and Love of God* takes most cogent form in the generating, "plastic" power that gyrates behind all forms of life, a "seminal force," as Swedenborg terms it. Swedenborg occasionally calls this universal will-to-life the "archeus," a term that indicates his indebtedness to Paracelsus as well as to the Cambridge Neoplatonists (and with them, the organizing "conatus" of Spinoza and Descartes). Within Swedenborg's emergent system of correspondences one can also detect traces of Paracelsus's older theory of analogical "signatures" present in nature, which had been amplified and disseminated by Jakob Böhme in the century prior to Swedenborg (though Swedenborg would later deny, perhaps misleadingly, that he had ever read any Böhme).[91] Apropos of Jane Bennett's pronounced use of Paracelsus to locate a form of vital materialism and proto-ecological sympathy, it is fitting to ask similar sorts of questions regarding Swedenborg's panentheistic instauration of the archeus in his *Worship and Love of God.*[92] Here, the archeus pushes life forms to replicate themselves and strive beyond their shapes. Poeticizing his earlier nebular hypothesis from the *Principia,* Swedenborg now describes the birth of the solar system in wholly natal terms (rather than the magnetic force of spirals and vortexes). The pregnant sun bursts forth his brood of planets, who circle "scattered like suckling masses near the burning bosom of their father and, as it were, at his teats" as they spiral outward into their respective orbits.[93] The planet Earth produces seeds and eggs of its own across its cooling surface that microcosmically mirror the egg of the sun and the production of planets. The animistic view of the Earth as a mother is concretized by Swedenborg—the earth's eggs hatch various animals who are "unfolded from their first wombs and eggs, and afterwards nourished as from a breast with the sweetest milk emanating from the florid bosom of her who gave them birth, and were brought even to that age when they were able to provide for themselves." Swedenborg's Earth in the *Worship* is fully an eighteenth-century baroque Gaia. Later, quadrupeds appear on Earth that sprout antlers as a sign of their arboreal descent from the egg-bearing tree branches.

When it comes to the birth of Adam, Swedenborg deviates from Genesis. Like the animals, Adam also grows out of a jewel-like vegetable egg hanging from the Tree of Life that is impregnated with the spirit of the Supreme Mind, and gestated with the dripping sap and juices of that tree.[94] Whereas Milton's *Paradise Lost* adheres fairly closely to the biblical account in Genesis as to why God created man, thus replicating the hierarchical theology that man must govern and rule (and name) the beasts of creation—a point in the translation of *Genesis* that ecocritics have always troubled—Swedenborg departs from biblical precedent and accounts for man's creation as one of aesthetic appreciation. Humans are made because there was no sentient being that could reflect, perceive, and synthesize sensations into thoughts about the wonder of the world.[95] This wonder is described as a "sacred astonishment" when Adam and Eve perceive the spirals, vortexes, and gyres as archetypal forms flowing beneath creation and its processes. The wonder also brings active enjoyment, and pleasure—the "very goal itself" of all life forms is to come in contact with feeling this (Divine) creative love as its point of origin: "to enjoy our Love is the veriest of life," an Intelligence tells Adam. "What is everything else but flying feathers, trifles, and dung! For into us He has transcribed Himself and His universal heaven." Here, God becomes a kind of inscription, an internalized language that is to be ecstatically apprehended. This aesthetic capacity of human consciousness thus becomes the primary means for the Divine Mind to return to itself (again revealing Swedenborg's debt to Neoplatonic theories of emanationism, its influx and efflux). From the moment of his emergence from the egg on the Tree of Life, Swedenborg makes Adam into a mini-Creator, an artist endowed with an inherent mimetic capacity: "[O]ur first-born, from the first time of his birth into the light of his world, acted like a delightful imitator, under the view and full control of the soul itself from which he was formed, and although ignorant of it as to the body, he nevertheless effigied and gesticulated." There are "mimic worlds constructed by minds" that have an organic rhythm and cycle, just like "those of the year and of the day." "Every mind constructs in effigy," an Intelligence tells Adam, "and as it were, builds some orb and world, in like manner as the Supreme Mind."[96] Here, as Jonsson has demonstrated, Swedenborg is invested in the hexaemeron tradition's use of Stoic notions of the *spermatikos logos*, where the mimetic power of the spoken (or written) word becomes a force generating matter, a law of evolution—human and Divine speech are active, creating principles, disseminating into the universe.[97]

The animate Earth in the *Worship and Love of God* that generates Adam and Eve is interconnected through this common, seminal force that establishes sets of relationships between all beings. Man is linked to creation via interlocking uses of an emanating nature, and Swedenborg's teleology here comes close to sounding like a more modern notion of the ecosystem. This generative vitalism "completes a chain of ends, in which every link, and every ligament of a link, may perform its gyre for the whole, and moreover that this concatenation may flow from perpetual uses . . . and thus may never cease."[98] The flow of forms here adumbrates Swedenborg's theory of influx that constitutes a theological reconciliation with both contemporary biological speculations on the originating principles of life (especially epigenesis) and more ancient Stoic notions of mystical forces that animate matter, creating what Bennett would call "vibrant matter." According to Jonsson, the resulting organic kinship between animate beings and humans in Swedenborg's *Worship and Love of God* recuperates "primitive examples of a belief in the deep affinity of organisms, and in that respect as distant relatives." The prose-poem, thus, harbors overt environmental ramifications; as Jonsson puts it, Swedenborg presents "a religious reverence for the life of nature. The human species is not allowed to ruthlessly tread down animals and vegetation or to look upon other organisms as machines."[99] The unceasing interconnectivity of Swedenborg's links and ligaments in nature offer a vertiginous decentering akin to the effects of what Deleuze and Guattari would call a "fiber," something that offers liberatory "lines of flight" for the self. As they put it in "Becoming Intense, Becoming Animal" from *A Thousand Plateaus:*

> A fiber stretches from a human to an animal, from a human or an animal to molecules, from molecules to particles, and so on to the imperceptible. Every

> fiber is a Universe fiber. A fiber strung across borderlines constitutes a line of flight or of deterritorialization. It is evident that the Anomalous, the Outsider, has several functions: not only does it border each multiplicity, of which it determines the temporary or local stability (with the highest number of dimensions possible under the circumstances), not only is it the precondition for the alliance necessary to becoming, but it also carries the transformations of becoming or crossings of multiplicities always farther down the line of flight.[100]

Ultimately, Swedenborg's theosophical coordinates prevent the kind of radical deterritorialization that Deleuze and Guattari's fibers presuppose, and the text of *Worship and Love of God* begins to collapse under the heavy weight of its own baroque symbolism and an increasingly elaborate system of didactic footnotes. The text's closing vision of burning lights and a beating heart is followed by a lengthy explication from Adam, who interprets every line of the vision's densely woven symbolism, with veritably every sentence interrupted by lengthy footnotes—thus, tautologically, presenting an exegetical commentary on an exegesis that is itself an imaginative revising of Genesis's original creation story. This tactic, of a visionary scene followed by an explanation, becomes a major rhetorical strategy in the subsequent theology. The dual structure of the *Worship and Love of God* itself, with its dense footnotes that mine below the passages on each page, also foreshadows the specific dialectics that Swedenborg's *Arcana Caelestia* would take up with Genesis and Exodus. The eight original Neolatin volumes of the *Arcana* (between twelve and fifteen volumes in the standard, modern English translations), consists of thousands and thousands of pages of detailed exegetical footnotes to Genesis and Exodus, line-by-line, often word-by-word. What pushes the *Arcana* beyond the boundaries of normal hermeneutics is not only the regular inclusion of Swedenborg's accounts of visionary things "seen and heard"—*ex auditis et visis,* as he refrains—in heaven and in hell, but his implicit suggestions that his revelation of the true internal sense of Judeo-Christian Scriptures constituted in itself an addition to God's Holy Word. Swedenborg's writings, in other words, were not supplemental, but could be seen as making a claim to be holy revelation in and of themselves: they constituted, Swedenborg suggested, the eschatological fulfillment of the Apocalypse foretold in the Book of Revelations. This was a line that even Milton hadn't dared to cross, though one could say Swedenborg's theological transgression further emboldened a number of his Romantic readers, from Blake, to Emerson, and to Whitman, who would continue conflating the religious into the literary, the poetic with the prophetic. It is to this visionary line, a kind of deterritorializing fiber, to which the next chapter will now turn.

2
Psychogeographies of Heaven and Hell

To the withered traditional church, yielding dry catechisms, Swedenborg let in nature again.

—Ralph Waldo Emerson

The writing of Swedenborg's unfinished *Worship and Love of God,* discussed at length in chapter 1, brackets his great spiritual turmoil and the transition from working as a natural philosopher with anatomical interests into a visionary exegete who traveled through heaven, hell, and other places.[1] Because of his scientific training, Swedenborg did not write like either the established Lutheran theologians of his day, or with the poetic extravagance or opacity of the mystics he is seen as closest in spirit to, such as Jakob Böhme. Instead, the later theology continues, more or less, the rational tone and expository rhetoric characteristic of his earlier Enlightenment scientific writing. In the "sordid whirling of infernal delights," Jorge Luis Borges writes, Swedenborg calmly stood, and "in flat Latin he catalogued / Whenless, whyless, eternal things."[2] For Borges (as for Czeslaw Miłosz, who calls Swedenborg's prose "soporific"), Swedenborg "prefers the exactitude and ultimately—why not say it?—the verbosity of the explorer or geographer exploring unknown lands."[3] Swedenborg's postscientific imaginary was fully cartographic; the opening sections of his *Heaven and Hell,* the little book that would later in the nineteenth century become his biggest success, begins with sections that carefully lay out how heaven is arranged in sequences of levels and planes—"anyone who does not know how the divine design is arranged in levels cannot grasp how the heavens are distinguished from one another," Swedenborg belabors, as he expounds on the differences between the natural, spiritual, and celestial realms.[4] No wonder, coming from the man who wrote some of the first modern geological descriptions of Scandinavia, that Swedenborg would feel so strongly about the proper abstract conceptualization of spiritual space. In the unpublished pages of his private journals that recorded the raw data of his visionary experiences, Swedenborg occasionally sketched small maps of locations (Jerusalem, Babylon, London) that he had seen in the spiritual world, with letter-coded cartographic keys explaining the movements of different groups of spirits through these spiritual spaces.[5]

Swedenborg's drifting and wandering through hellish caverns and beautiful, heavenly gardens can even be seen as a crucial eighteenth-century precursor to Guy Debord's later "psychogeography," a trope of peripatetic writing-as-walking that harbors a potential for radical, visionary insight.[6] Elements of Swedenborg's thought are traceable in both the roots and later fruits of Debordian psychogeography. Debord's Situationist aesthetics drew deeply on the earlier work of Walter Benjamin, particularly Benjamin's long sequences on the wandering urban flâneur in his *Arcades Project* (1927–40)—which, in turn, quotes copiously from the French Symbolist Charles Baudelaire's writings.[7] As this chapter takes up, Baudelaire channeled Swedenborgian correspondence theory directly into his "Correspondances," one of the central poems from *Les Fleurs du mal* (*The Flowers of Evil*, 1857), a foundational text for Modernist poetics. In more recent years, the British writer Iain Sinclair has revitalized Debord's psychography to (re)read contemporary London urban space, uncovering an occult and esoteric geography that incorporates the earlier walking patterns and writing habits of both William Blake and Swedenborg himself. For Sinclair, these obscured and forgotten networks of spiritual walking-as-seeing are powerful loci of critique and resistance to the deadening forces of secular modernity and late-stage capitalism.[8] Swedenborg's later theological writings, particularly the posthumous manuscripts of spiritual experiences that more freely play with language and space, could provide a kind of template for these poetic projects of urban reenchantment, the blueprints for building a utopic New Jerusalem "in England's green and pleasant land," as Blake famously put it in 1804.[9] Deeply encoded in the DNA of the urban flâneur, or lurking in the periphery of these alternative cartographies of the psychogeographer, one can find fragments of Swedenborgian theology.

David Dunér interprets Swedenborg's later mysticism as a continuation of deep cognitive patterns that were already very much present in the earlier geology and geography: only now, "as a visionary he built up a metaphorical system, a doctrine of correspondences and a spiritual world, all predicated upon metaphorical thought. Swedenborg created his own work of art, into which he finally stepped, to disappear out of the prison of reality. The work of art was confused with reality, or perhaps rather became his reality, the artist's dream of becoming one with his work."[10] Friedemann Stengel, by contrast, sees the theology as the inevitable outcome of a "visionary empiricism" that turned to heavenly and spiritual matters, a kind of response to the pressure exerted on spiritual truth to have an experiential, materialized foundation.[11] Whatever the multifarious causes, Swedenborg's great shift is entangled in an enormous struggle within and through language, as the copious pages of his private journals and dream diaries from the 1740s readily attest. "All was such that I cannot describe" is an apophatic refrain that threads these thousands of pages (published only posthumously) together.

In reading through these pages, one can also chart Swedenborg losing his sense of self. After a sharp, vivid dream, he writes in his journal how he has "finally come into such a state that I know nothing, and that all *preconcepta judicia* [preconceived opinions, biases] have been taken away from me, which is the beginning of learning, namely, that one first must become a child, and then be nurtured into knowledge, as is now, I think, finally happening to me."[12] A few years later, he characteristically jots down how a person "begins to become wise who knows little, virtually nothing from himself as the source. This is the same to say that he who is nothing is something."[13] Rather than such statements being merely incidental platitudes, Swedenborg integrated this feeling of "nothing as something" into his larger theological project, and his particular hermeneutics concerning the Book of Genesis. The words there that Swedenborg struggled to exegete were not in any way a literal or historical account of the creation of matter and the world, but rather a complex allegory of spiritual growth, suffering, and psychological regeneration. Humans suffered because of their egotistical love of themselves and their attachments to things (the "love of the self," *amor sui,* and the "love of the world," *amor mundi,* as Swedenborg puts it), when in actuality, the self was simply an empty vessel, a nothing in-and-of itself outside of divine presence. The consequent lack of there being "a there there," in Swedenborg's ontology, and his positive valuation of a state of nothingness, comes closer to Buddhist notions of Śūnyatā than anything remotely familiar within the Christian theological discourses of Swedenborg's day (excepting other heterodox mystics such as Meister Eckhardt and, less so, Jakob Böhme). D. T. Suzuki is but one of several modern Buddhist thinkers who consequently hailed Swedenborg as a "Buddha of the

North" for the European West.[14] Miłosz saw this buddhistic aspect of Swedenborg as specifically undoing the Cartesian bifurcation of mind and body, spirit and world, rendering Swedenborg "an exception among European thinkers" for his radical deconstruction of the individual self.[15] Swedenborg's undoing of what we tend to call the formation of the singular "modern subject" during the Enlightenment presents an untapped resource for certain strands of the new materialisms that have turned "east" for their ontology and, on occasion, their calls for new forms of ecospirituality.[16] Swedenborg's Śūnyatā-like observations on the self emerged out of his wrestling with language: from an attempt to extract a spiritual hermeneutics from the correspondential Hebrew phrases of the Bible, on the one hand, and to find adequate Latin or Swedish words that could capture his intensely wordless, inner spiritual experiences, on the other.

Though a series of strong dreams and visions of Christ in 1745 had confirmed Swedenborg's sense that he was being commanded to *do* something, the precise nature of his holy commission remained obscure. It wasn't until the appearance of the first volume of the *Arcana* in 1749, four years after the *Worship and Love of God* had petered out and the period of turmoil in the journal of dreams, that Swedenborg finally seems to have located a writing and exegetical style in which he was confident.

"Very, Very Strange Correspondences"

Swedenborg's doctrine of correspondences, heralded in his earlier scientific work on the lungs, had now found anchor in a reenchantment of biblical text, turning its words into the Word. During this three-year interim, Swedenborg's reading and writing patterns underwent significant alteration (he also, as has been noted, began gardening in earnest at his new property on Södermalm). Swedenborg forsook his close readings in philosophy and biblical commentary, turning to the original Hebrew and Greek Judeo-Christian biblical texts in an attempt to unearth their correspondential structures and original spiritual internal sense. His painstaking etymologies of Hebrew words in Genesis as he searched for an authentic bedrock of celestial meaning were shaped by his research and interest in hieroglyphics and the materiality of a visual representation of thought. He may have also been impacted by kabbalistic traditions that had highly developed esoteric theories on the relationship between the form and sound of Hebrew characters and the content of secret spiritual wisdom that lay in their combination into various compounds and words.

The question over Swedenborg's relationship to Jewish sources remains vexed, however, and continues to be a place of fruitful debate among contemporary scholarship on Swedenborg.[17] Given the specifically aural and sonorous dimensions of enchantment as a mode or affect—Gilles Deleuze and Félix Guattari, Jane Bennett, and David Abram have all emphasized the *chanting* aspects of enchantment—it is tempting to conjecture Swedenborg repetitiously uttering Hebrew syllables in a meditative manner as he worked on a massive personal index of the symbols and correspondential themes he was unearthing in the Bible, reading the texts again and again, as he simultaneously documented his ongoing mystical experiences in his private journals.[18] Some of these activities evidently converged with one another in the opening pages of Swedenborg's diaries of spiritual experience, which ostensibly begin with a biblical index cataloguing scriptural references to *Zea*, *Zelus*, and *Zona* (Latin for "corn," "jealousy," and "belt"), before proceeding to provide numbered accounts of spiritual visions, which, shortly thereafter, is interspersed with a page of Hebrew verb conjugations as Swedenborg works out the sound and spelling of various verbs. Whatever the intent, the heterogeneous miscellany of these opening pages suggest how Swedenborg's diary functioned more as a commonplace book, and less as a systematic or coherent theological exposition of the spiritual world (as it has sometimes been read by Swedenborgian church movements). The Hebrew content that appears in the diary may just indicate that Swedenborg was merely doing his homework, trying to practice conjugating his verbs in a new (biblical language), in between episodes of visionary experience—and not evidence, per se, of a kabbalistic kind of reenchantment. Either way, this repetitive activity of translation and exegesis—shuttling back and forth between apophatic experiences of the self and the textual work of biblical hermeneutics—marks a Deleuzian reterritorialization of Hebrew biblical text. As Deleuze and Guattari observe, the sonic refrain is essential in creating "territorial assemblages," and, even, the delineations of the aesthetic.[19] Jesper Swedberg had also insisted to his students at Uppsala (including his then fifteen-year-old son Emanuel) how "the Bible must be your handbook, *you must be at home in it*," and now, years later, Swedenborg was doing his utmost to make a (new) home in the (old) Bible, creating a dazzling system of words and correspondences through an exegetical method that had been sharpened by decades of his scientific training.[20] A single correspondence selected from one of Swedenborg's indexes radiates off into multiple directions for interpretation and meaning. It is this liberating possibility, away from the singularity of a literalist approach to biblical texts, that lets Joan Richardson write of Swedenborg's world of words as a form of "linguistic cosmology."[21]

Swedenborg's "science of correspondences" was thus both a hermeneutic for exegeting sacred Scripture as well as a way of perceiving the phenomena of the external world, where "absolutely everything in nature, from the smallest to the greatest, is a correspondence." While Swedenborg posited that true understanding of these spiritual semiotics encoded in nature was largely a lost art to the contemporary world (though he intriguingly suggests that it perhaps continued as an ancient practice in places far outside the pale of Christendom, in Africa and "Greater Tartaria," or Mongolia), the suggestion of its presence existed in tension with Swedenborg's attempts to anchor the symbolic system of reference to sacred Scripture. His "language of things," the wish for "words to be one" with objects, as his later American disciple Sampson Reed so colorfully puts it, never fully merges with the things-in-themselves. Such a substrate of meaning is perhaps the apophatic space beyond words and representation, and Swedenborg simply remains too much of a Cartesian—or a "melancholy Dualist," to evoke Timothy Morton's ~~pantheism~~ that was discussed in the last chapter—to allow such a collapse to happen. As Wouter Hanegraaff puts it, Swedenborg was pushed to develop his theory of correspondences "*not* because it unites mind and nature, but for precisely the contrary reason: because it solved (for him) the Cartesian dilemma. It enabled him to retain a fundamental distinction between spirit and nature (modeled on the distinction between Descartes' *res cogitans* and *res extensa*); to continue seeing spirit as the active cause and nature as the passive substance 'acted upon' by spirit; but to explain that relationship without having to assume the existence of mechanisms of instrumental causality."[22] In such a schema where correspondences lay behind every fact of a nonmechanized, vitalistic nature, all the natural truths of science were translatable into spiritual precepts, and nature became electrified with revelatory potential. For Ralph Waldo Emerson and others, this facilitated the displacement of the dynamics of conversion into a natural (and thus unecclesiastical) space: "in the woods, we return to reason and faith," as Emerson puts it in his *Nature*. Not in the church.

As a system of signs, Swedenborg's correspondences share more with the subsequent semiotics that came to be developed by Charles Sanders Peirce than they do with the predominant strain of linguistics traceable back to Ferdinand de Saussure. Saussure—and, one could argue, the waves of structuralist and poststructuralist lines that followed in his wake, including certain kinds of deconstruction—tended to view semiosis as more or less occurring wholly, exclusively, within human articulate language. There is little maneuvering for the other-than-human world between either Saussure's signifier and signified, or Jacques Derrida's critiques of logocentrism—his oft-misunderstood contention that there is "no world outside the text."[23] Peirce, by contrast, saw semiosis as inhering to the fabric of all matter, everywhere—a starting point that has made his work of seminal importance for biosemiotics and some of the new materialists.[24] Peirce described his sense of this as a quasi-mystical feeling:

> It seems a strange thing, when one comes to ponder over it, that a sign should leave its interpreter to supply a part of its meaning: but the explanation of the phenomenon lies in the fact that this entire universe,—not merely the universe of existants, but all that wider universe, embracing the universe of existants as a part, the universe we are all accustomed to refer to as 'the truth'—that all this universe is perfused with signs, if it is not composed exclusively of signs.[25]

Or, as Swedenborg puts it once again in the "Language of Heaven" chapter from *Heaven and Hell:* "in a word, absolutely everything in nature, from the smallest to the greatest, is a correspondence."[26]

To align Swedenborg's allegorical correspondences with Peirce's later American semiotics at this juncture is perhaps not as odd or dissonant as it might initially seem. Peirce was fundamentally affected by his close affiliation with Henry James Sr., and James's idiosyncratic brand of nineteenth-century Swedenborgianism. Andrew Taylor reads James Sr.'s ideas as imbuing Peirce's philosophical preoccupations with fallibilism, and the productive, vague spaces of uncertainty and skepticism; Donna Orange and Paul Jerome Croce see James's Swedenborgian theodicy—his reconciliation of the concrete facts of evil and suffering with a notion of the Divinity—as further coloring Peirce's own "neglected" arguments for God.[27] While there is not space here to more fully draw out this rich sort of "ecotone" that lies between Peirce and Swedenborg, mediated by the various Jameses, it suffices to observe that even if there is no direct causative correlation to draw between Peirce's theory of signs and Swedenborg's correspondences, it seems no mere coincidence but a fact of the "survival of the fittest" in the genealogical history of ideas that Peirce's ideas came to germinate in both a general Transcendentalist milieu that was steeped in an Emersonian nature-as-language, and in a specific discursive space of the James family dinner table, where Henry James Sr. demonstratively expounded and adapted Swedenborgian ideas to his own eclectic philosophy and system of beliefs. These were both generative contexts.

Of course, Swedenborg never systematically elaborates his sign theory to the same rigorous extent that Peirce developed his typologies of icons, indices, and symbols. And other mystics and figures that Peirce was aware of certainly also resort to a similar classic "book of nature" trope, from Jakob Böhme's Paracelsian "signatures" to the elaborate typologies developed by the Puritan philosopher and theologian Jonathan Edwards. This is what Hans Blumenberg has called the evolution of the (theological) idea of the "Lesbarkeit der Welt," the readability of the world.[28] Quite different from Böhme or Edwards, however, Swedenborg consistently refers to his sign-doctrine as a "science," and his concept, unlike Edwards's typologies that hew closely to a standard Protestant *sensus literalis* of sacred Scripture, originated out of the specific anatomical investigations into the communicative nature of matter in the human body. Swedenborg's speculation on kidney functions in *The Animal Kingdom*, as discussed in chapter 1, prompt his first articulation of this "science of sciences" about signs; it exemplifies the "seeking of form in facts," to paraphrase Alfred North Whitehead, that was latent in much Platonic thought.[29] It was even the vagueness within Swedenborg's correspondential system, its unarticulated tension between sacred Scripture and the ongoing revelatory power of nature, that led to its flexible adaptations within later nineteenth-century poetics.

Among the most well-known poetic instantiations of Swedenborg's correspondences is Charles Baudelaire's Symbolist sonnet "Correspondances" from *Les Fleurs du mal*, which directly stemmed from Baudelaire's encounter with Swedenborgian ideas.[30] "Correspondances" reads as a Continental analogue to the American language of things that this book is exploring; Baudelaire characteristically transforms Swedenborg's original movement between Scripture and the Book of Nature into a movement between artistic forms, between different sorts of representation. The resulting synesthesia—also, like Sampson Reed, a wish for words "to be one with things"—is a striking, sensual intensity that suggests the substrate of pure being that underwrites all representation itself. Correspondential words are like "perfumes," Baudelaire writes at the poem's end, that are

> fresh as children's flesh
> Soft as oboes, green as meadows
> And others, corrupted, rich, triumphant
> Possessing the diffusion of infinite things
> Like amber, musk, incense and aromatic resin
> Chanting the ecstasies of spirit and senses.[31]

This "ecstatic chanting" famously conflates the olfactory with touch, sound, and sight. Gilles Deleuze may have even had Baudelaire's particular synesthesia in mind when he extemporized in one of his seminars how "between a philosophical concept, a painted line and a musical sonorous bloc, resonances emerge, *very, very strange correspondences* that one shouldn't even theorize, I think, and which I would prefer to call 'affective' . . . these are privileged moments."[32] Bernd Herzogenrath usefully glosses on Deleuze's remark here on how this privileging is "an affect where thought and sensation merge into a very specific way of doing thinking beyond representation and categorization, a moment that might be called 'contemplation.'"[33] This appropriation of a word and affect ostensibly associated with religious or spiritual exercises, such as "meditative contemplation," suggests one avenue that Deleuze might be approached in a postsecular kind of way; to contemplate the correspondences of nature—this living "forest of symbols," as Baudelaire phrases it in his famous poem, that "watch man with intimate eyes"—could become a mode of reenchantment.

Interstellar Ecology

Swedenborg's own theological contemplation of the cosmos departed from the nebular hypothesis of the earlier *Principia,* using correspondential analogies to expand the micro to the furthest limits of the macro. All good souls of the universe, when viewed as a composite single organism, took on the shape of an enormous human being, the *Maximus Homo,* the "universal human" (or "Grand Man," as it was translated in the nineteenth century). The figure bears obvious parallels to both the realms of political theory (Hobbes's *Leviathan* had appeared in 1688, the year of Swedenborg's birth) and more esoteric kinds of cosmology, such as the notion of the Adam Kadmon in kabbalistic thought (even as Swedenborg's relation to esoteric Judaism remains an open question). Paralleling the Kabbalah, Swedenborg earthily embeds the *Maximus Homo* to each and every living human body: "[W]hat is wonderful and hitherto unknown, all parts of the human body correspond to societies in heaven. And therefore it is sometimes said that some societies belong to the province of the head, some to the province of the eye, others to that of the chest, and so on."[34] The mind is surrounded by a swarm of spiritual "societies" tied to particular anatomical organs of the corporate *Maximus Homo,* whose collective voices form the background "white noise" of our consciousness. When we track the gradual appearance of this idea through the pages of Swedenborg's unpublished diaries, these voices from the Other(s) displace Swedenborg's sense of himself (as by this point, he was coming to the conviction that our sense of ourselves—what he calls in Latin the "proprium"—as discrete individual units with subjectivity and agency was, in fact, the grand illusion). Swedenborg's angels "assemble" (his word) into groups and aggregates—the organs of the *Maximus Homo* that are respectively constituted by so many individual cells, acting in concert—and they then act as one on the flow of our individual thoughts and the actual sensations within the corresponding individual organs of our physical bodies.[35] The concept is simultaneously enfleshed in the matter of bodies, at the same time it dematerializes into discourse: the language of our thoughts is part of a vast other kind of collective corporeality.

In a late, remarkable essay by Emerson that further elaborated his sense of language as a living, organic entity that was shaped by processes of natural selection and evolution, he returned to this somewhat arcane idea in Swedenborg, and used it as a fulcrum to turn to a larger point about the unsettling, depersonalizing power of language. "Swedenborg threw a formidable theory into the world," Emerson writes, alluding to the *Maximus Homo:*

> that every soul existed in a society of souls, from which all its thoughts passed into it, as the blood of the mother circulates in her unborn child; and he noticed that, when in his bed,—alternately sleeping and waking,—sleeping, he was surrounded by person disputing and offering opinions on one side and on the other side of a proposition; waking, the like suggestions occurred for and against the proposition as his own thoughts; sleeping again, he saw and heard the speakers as before: and this as often as he slept or waked. . . . Language is a city, to the building of which every human being brought a stone; yet he is no more to be credited with the grand result than the aceleph which adds a cell to the coral reef which is the basis of the continent.[36]

As medievally occult, then, as Swedenborg's *Maximus Homo* might seem to be, it nevertheless helped catalyze Emerson's perception of language as both a spiritual office and a natural being in itself in which individual humans were displaced, becoming something like "flies, aphides, gnats, and innumerable parasites," in Emerson's words, blindly sucking up various content. Emerson's creative reading of Swedenborg suggests how the *Maximus Homo* might be further seen as an imaginative instantiation of another initially bizarre-sounding concept: Deleuze and Guattari's so-called Body-without-Organs, or BwO: a cryptic, "thick" description that has come to be an important locus for a number of new materialist thinkers. For Deleuze and Guattari, the BwO is a deeper substrate of reality underlying the formation of separate entities, organs, sensing bodies—it is a constant "plane of immanence" grounded in the generative earth.[37] The goal of philosophy, according to Deleuze and Guattari, aims to liberate subjects (back) to the BwO, or becoming a BwO. To situate the BwO with Swedenborg in any kind of constellation, however, ostensibly requires either detheologizing Swedenborg or mystifying (in the worst sense) Deleuze and Guattari, as their shared philosophical project is deeply suspicious of the transcendental, unifying ontotheology that someone like Swedenborg so thoroughly evinces. Representational thinking, analogy, and correspondence are precisely the terms that Deleuze and Guattari, particularly in *A Thousand Plateaus,* attempt to implode through various alternative figurations, such as the rhizome and fold. As significant as these differences are, nevertheless, the net effect and result of Swedenborg's *Maximus Homo* achieves surprisingly similar ends as Deleuze and Guattari's BwO, which proleptically charts a project of liberation: Swedenborg's "MaHo," this section proposes, can be read as a kind of theological inverse of the BwO.

Swedenborg's *Maximus Homo* is first fully articulated in his multivolume exegesis of Genesis, the *Arcana Caelestia,* but is perhaps most recurrent in the later *Earthlike Bodies Called Planets* (1758), an account of Swedenborg's interstellar journeys to visit spirits on the planets of this solar system, and beyond. This work is also Swedenborg's clearest participation in a burgeoning eighteenth-century genre of utopian literature, following in the tradition of Thomas More and Jonathan Swift's *Gulliver's Travels* (1726), where the no-space of distant lands veils a trenchant critique of a close-at-hand present. As Swedenborg's spirit travels out into space, his encounters with extraterrestrials locate planets and peoples, on the one hand, within the infinite macrocosmic body of the *Maximus Homo.* Spirits from Venus live in the memory; those from Jupiter reside in the testicles; many from Mars congregate in the longitudinal sinus of the brain. On the other hand, the encounters and journeys are intimately mapped onto the spaces of Swedenborg's own physical body. He feels an "influx" on his spleen when conversing with denizens of a distant planet.[38] He watches spirits from Mercury congeal into a globe, and roll along toward the inhabitants of Venus, which causes Swedenborg to feel "a remarkable change in the brain."[39] The space of the text is deeply allegorical; Anders Hallengren usefully suggests that the proper framework for locating the planets and aliens in the *Earthlike Bodies Called Planets* is not early eighteenth-century astronomy, but rather premodern astrology.[40] A medieval medical context is suggested by this grafting onto specific organs of Swedenborg's individual body, often in ways that disconcertingly "deterritorialize" out into the stars. "Once while I was writing concerning things to come," he recollects,

> some spirits from Mercury were at a distance, so that they could not look into those things from my memory, because I was not willing to read them in their presence, they were very indignant, and contrary to their usual behavior, they were desirous to inveigh against me, saying that I was one of the worst of men, and the like; and that they might give proof of their resentment, they caused a

kind of contraction attended with pain on the right side of my head even to the ear.[41]

Deleuze and Guattari's BwO is a complex of metaphors and observations meant to stimulate its readers toward liberatory modes of affect and feeling that disintegrate the organicism of the body as a singular holistic entity—a holism and organicism that Deleuze and Guattari see as suspect when the body is metaphorically transferred to uphold and corroborate the modern liberal subject, the individual and autonomous self, and all its appurtenances of (state and social) control. Most fully articulated in *A Thousand Plateaus,* the dense collocation of aphorisms and quotes about the BwO opens up a program for the reader to intensify sense (especially pleasure), to get at the "ontological unity of substance," a feeling of the "continuum of all substances in intensity and of all intensities in substance."[42] This earthy turn outward to sense deliberately keys the monadism of the seventeenth and eighteenth centuries that also shaped Swedenborg—"all BwO's pay homage to Spinoza," as Deleuze and Guattari lightly put it—and it is one reason why the BwO has become an important figure in the conceptual toolbox of new materialist revitalizations of Spinoza.[43] The BwO further recuperates and frequently tropes on the older Renaissance concept of the cosmic egg that had been so generative for Swedenborg's nebular hypothesis. The BwO is, admittedly, hostile to the theological: "[T]he enemy is the organism . . . the organic organization of the organs. *The judgement of God,* the system of the judgement of God, the theological system, is precisely the operation of He who makes an organism, an organization of organs called the organism, because He cannot bear the BwO, because He pursues it and rips it apart so He can be first, and have the organism be first."[44]

Nevertheless, if we carefully parse the ways that Swedenborg's writing on the *Maximus Homo* and in the *Earthlike Bodies Called Planets* slips and slides away from a singular self, as other spirits and voices come to speak through Swedenborg in the very act of writing, and further observe how these displacements are so often materialized in specific, concrete places of feeling in Swedenborg's body (pains in his stomach, influxes in the brain), Swedenborg's micro-macro cosmic dilations end up instantiating the dismantling effects that Deleuze and Guattari's BwO sets out to achieve: "[D]ismantling the organism has never meant killing yourself, but rather opening the body to connections that presuppose an entire assemblage, circuits, conjunctions, levels and thresholds, passages and distributions of intensity, and territories and deterritorializations measured with the craft of a surveyor. . . . [H]ow can we unhook ourselves from the points of subjectification that secure us, nail us down to a dominant reality?"[45] Swedenborg answered this question by imaginatively embarking on a voyage to the stars that also folded around the defamiliarizing interstices of his own body.

Swedenborg's cosmic *Maximus Homo* remains, in the end, a classic *imago dei*, rendering it ultimately incompatible with a Deleuzian BwO. As the corporate body of the cosmos that refracted the ideal order of God, Swedenborg's *Maximus Homo* has been read by Constantin Rauer and others as even abetting a proto-fascistic order—a reading that would seem to justify the stringent antitheism of the BwO, and Deleuze and Guattari's concern that such holistic organicism can constitute a "power takeover" that flattens multiplicity, erasing difference. Rauer traces a Platonic strand epitomized by Swedenborg within European Continental Romantic traditions, drawing a line from Swedenborg's organic cosmology on to Johann Wolfgang von Goethe, Johann Kaspar Lavater, Friedrich Wilhelm Joseph Schelling, and G. W. F. Hegel, and then later into more explicitly totalitarian contexts, from Hans Driesch, Rudolf Steiner, and Franz Hartmann to the enfleshed "Blut-und-Boden" ideologies of Adolf Hitler. The vast "puppetry" behind Swedenborg's platonizing *Maximus Homo*, Rauer claims, lies at the kernel of Nazi biopolitics.[46]

A key part of Rauer's argument is juxtaposing Swedenborg against Immanuel Kant, and reading Kant as essentially anti-Platonic (a bit of philosophical good cop, bad cop). As Friedemann Stengel has shown, whatever the case of Rauer's Lacanian thesis—his claim that Swedenborg exhibits characteristic "schizography"—he nevertheless doesn't adequately address the complex, sustained reaction that Kant actually had to Swedenborg's ideas.[47] A careful reading of Swedenborg's *Earthlike Bodies Called Planets*—which may, in turn, refract Swedenborg's awareness of Kant's own treatise on the solar system and the planets that had appeared three years earlier, in 1755—would also seem to belie Rauer's sweeping generalizations. Swedenborg's planetary text intimates forms of ecological cosmopolitanism, not political authoritarianism: kinds of belonging that go beyond the local and national and, I argue, even the recognizably human, as Swedenborg visits stranger and more unknown corners of the universe along his astral journey. The entelechy of Swedenborg's *Maximus Homo* is not the confines of fascism, but the openness of unknown outer space.

As a piece of utopian literature, *Earthlike Bodies Called Planets* express skepticism toward the concept of private property and ownership (which derive from, according to Swedenborg, a love of possessing things and our selfish egos—the source of all social evil and suffering). The text's celebration of the diversity and difference in the universe further unfolds an anticolonialist critique, as the pastoral visions of life on other planets largely read as a fable of premodern enchantment. Swedenborg's extraterrestrials usually live nomadic or semi-settled lifestyles, abstaining from eating meat and foraging for plants for food. They organize tribally, associating as families, with no written law or systems of government as "it was a thing altogether unknown for one person to enrich himself from the goods of another," as the testicular spirits of Jupiter inform Swedenborg: "No one ever covets the goods of another; that it never enters their minds to desire the possessions of another, still less to obtain them fraudulently, and least of all to break in and plunder."[48] On a remote world far outside our solar system, Swedenborg encounters a society that was being terrorized by monks from planet Earth, who were attempting to force conversions to Christianity and make the extraterrestrials "live together in society." The aliens are warned (ostensibly by Swedenborg) that these monks from Earth only wish to do so that "they might rule and have dominion over them . . . and make them slaves."[49] Swedenborg's writing here may have refracted his awareness of New World atrocities committed by Spanish colonists, as famously recounted by Bartolomé de Las Casas in 1552—Swedenborg takes note elsewhere of "the butchery of the Spaniards in America, where they tracked down the heathen and cruelly tore them to pieces by means of dogs."[50]

The diversity of life on other planets extends to the human form, which pushes beyond normative categories of Homo sapiens. Spirits on Mercury are unwilling to appear as earthly humans, and prefer to rather take shape "as crystalline globes." They speak in undulatory waves that physically and synesthetically affect Swedenborg's left eye. On Jupiter, the inhabitants have enlarged faces, wear tight-fitting blue clothing, and walk like simians—"as they go along, they assist themselves on their feet . . . which is done suddenly."[51] On Venus, a race of large, rapine giants keep cattle (husbandry being an encroaching sign of *civitas*, conspicuous among all the other planetary pastorals of the text), and these giants horrifically delight "in eating their plunder." Men on the moon do not speak and breath from the lungs like other varieties of humans, but belch out "thunderous eructations" of speech and sound from their abdomens.

One of the most interesting and detailed descriptions of extraterrestrials occurs in Swedenborg's descriptions of a "third earth in the starry heaven," far beyond the solar system. Today it reads like an anticipation of J. R. R. Tolkien's description of the elven village Lothlorien from *The Lord of the Rings* (another indication of how the utopianism of *Earthlike Bodies Called Planets* dovetails with the parallel emergence of the speculations afforded by eighteenth-century science fiction). Far exceeding the skill and beauty of earthly palaces made of marble and stone, Swedenborg writes, the inhabitants of this third earth excelled in creating living, arboreal architecture. By shaping and pruning enormous trees as they grew, these extraterrestrials caused the branches to intertwine, creating living floors and walls, and staircases elevated high above the ground. In the apertures between the branches that form these interconnected holy sanctuaries, prisms were hung that refracted hues of blue and orange light. As we will see in chapter 4, these otherworldly descriptions may have colored the development of a distinctly environmental architectural style in the San Francisco Swedenborgian church in the 1890s, a space imbricated in the emergence of the Sierra Club and John Muir's preservationist politics.

Outlandish as all this material in *Earthlike Bodies Called Planets* is, the plethora of almost-anthropological detail in the text serves to underscore Swedenborg's repeated assertion that "where there is a planet, there are humans; for the human is the end for which every earth is created."[52] This teleological peopling of the universe was by no means uncommon in the eighteenth century; Kant had undertaken similar speculations in 1755 (if without Swedenborg's visionary empirical claims), and as late as 1779, Wilhelm Herschel was still attempting to prove to the British Royal Society in London that the architecture of subterranean alien cities was discernable on the surface of the moon, visible through a telescope. Herschel's subsequent astonishing discovery of Uranus, in 1781, was enabled by his optimistic feeling that the cosmos was a vast home for familiar life, as Richard Holmes has so expertly detailed.[53] In its own queer way, Swedenborg's earlier *Earthlike Bodies* comes from a similar place of affirmative feeling, of a philosophical desire to be at home, if not in the world so much as in the universe. If we recall Swedenborg's parallel observation that "there is something of the human form in all the universe," his not-so-apodictic insistence that "where there is a planet, there are humans" might be reformulated as "where there is a planet, there is some kind of life"—a vitalizing force that causes humans and nonhumans alike to be something (rather than nothing) on an earth. Like Herschel's late cosmology that imagined the universe as a giant interconnected organism, Swedenborg's incorporating figure of the *Maximus Homo* in *Earthlike Bodies Called Planets* seems less an incubator for twentieth-century totalitarianism and more of an anticipation of how picturing planet Earth within the wider macrocosm could productively galvanize environmental imaginaries, as later famously happened around the reception of the "blue marble" image of the earth that was taken by the *Apollo 17* space mission in 1972. As Ursula Heise and Finis Dunaway have respectively shown, the famous astronautical photograph of a jewel-like Earth, floating in black space, catalyzed environmental discourse in the 1970s, influencing both popular movements behind the first Earth Days as well as thinkers such as James

Lovelock and Arne Naess.[54] Whereas this image caused a collective sense of the singular fragility of Earth and its ecosystems, surrounded by the black vastness of outer space, Swedenborg in his *Earthlike Bodies* expresses characteristic Enlightenment optimism that was equally shared by Herschel and others who had speculated on the viability of extraterrestrial existence: where there were worlds, there were certainly people. Nevertheless, the cosmological vastness and plural forms of life in *Earthlike Bodies Called Planets* causes an inevitable diminution of the geocentricism of planet Earth, furthering a displacement that had begun with Copernicus, a "smallness" that continues to operate in the constructed preciousness of the blue marble image from 1972. Swedenborg would have read in John Milton's *Paradise Lost*—a poem that deeply colors Swedenborg's *Worship and Love of God*—the famous scene where Adam looks up and wonders at the celestial bodies above the Garden of Eden, if they, too, are suns with revolving planets and various inhabitants. The angel instructor Raphael quickly admonishes Adam for his curiosity, and enjoins him to

> Dream not of other worlds, what creatures there
> Live, in what state, condition, or degree;
> Content that thus far hath been revealed
> Not of earth only, but of highest Heaven.[55]

Only a century after Milton, emboldened by an empiricism and his theologically driven cosmology, the angelic spirits would seem to have changed their tune, and were now freely instructing Swedenborg about the "millions" of earths in the universe. "What would the human race," writes Swedenborg, "be for the infinite Creator, for whom a thousand earths, nay tens of thousands, would not be enough?"[56] Our Earth was but one of many in the starry heavens.

Epigenesist Influx

In Swedenborg's view, the entire cosmos and its scintillating points of correspondence were held together by influx, Swedenborg's variant on Neoplatonic emanation theories. The mind-body-matter problem, made so acute by Descartes, and as refracted in Swedenborg's copious notes and quotations of Leibnitz, Spinoza, and Malebranche in his 1740s *Philosopher's Notebook*, remained a serious point of preoccupation well into Swedenborg's theological period. Quite late in his publishing career, at the age of eighty-one, Swedenborg felt the need to address the issue singularly, and he brought out a small volume, the *Interaction between Soul and Body* (1769). In the early nineteenth century this was occasionally translated as *A Theosophic Lucubration on the Nature of Influx*, a copy of which was purchased by William Emerson, Ralph Waldo Emerson's father. *Influx* was probably the young Emerson's first encounter with Swedenborg's theology.

The little book sets out how Swedenborg's influx was an outflow of divine love animating the forms (the wisdom) of all created matter. Nature appears dead and lifeless without this perpetual inflowing power permeating all. Swedenborg's influx was not only Platonic; his theological reflections in *Interaction* have recourse back to the great biological debates about preformation and epigenesis his earlier science had been so preoccupied with. In his *Divine Love and Wisdom,* Swedenborg had recounted overhearing the spirits of two deceased presidents of the British Royal Society, Sir Hans Sloane and Martin Folkes, hashing out and debating the origins of seeds and eggs in the spiritual world (Swedenborg had known both men personally, and associated with them during his earlier sojourns in England). Sloane ends up being convinced—parroting Swedenborg's own position, of course—that the propagation of all animals and plants is but "a flow" of constant spirit into nature. Sloane "went on to say," Swedenborg reports, "that if he had known what he now knew about the spiritual world, he would not have attributed anything to nature except the service it provides to spiritual reality from God."[57] Though there are vestiges of the *spermatikos logos* and *archeus* that had driven Swedenborg's earlier emanation theories in his scientific writings, now in the theology, as in certain schools of Sufi mystics or the Kabbalah, Swedenborg places a concept of divine love at the teleological center, a love that is perpetually seeking to reunite created matter with itself: the world is but a continual dynamic force whose goal is "a striving to return to seed."[58] Forms and species accordingly propel upward into each other. Minerals aspire to be plants, and in all plant life "there is an energy towards movement" that approximates animal life.[59] "There is a ladder of all created things," writes Swedenborg, an interconnectedness that should "stun" the mind with wonder. Putting Swedenborg's theism aside, this permeability between species and forms can be seen as broadly anticipating later nineteenth-century theories on evolution and natural matter, especially Jean-Baptiste Lamarck's pioneering work on the transmutation of species. An incipient "trans-corporeality" lurks within Swedenborg's influx, one that might have remained within the orthodoxy of Christian Neoplatonic beliefs in a definitive, gradated scale of Being were not Swedenborg's concept so thoroughly informed by his lifetime spent studying scientific phenomena.[60] Years earlier, Swedenborg had offered the astonishing speculation in the *Principia* that the beginnings of forests might have been in the accidental flaking off of minerals from rocks onto beds of lichen and moss—a union of active and passive substances that began to produce new life forms, the trees (we could also recall here the charming poetic detail, discussed in chapter 1, in the creational prose-poem *The Worship and Love of God*, where quadrupeds like deer sprout branching antlers as signs of their arboreal origins from the Tree of Life).

Swedenborg's influx shows how his Cartesian mechanism could be fundamentally augmented—softened, organicized—by a recuperation of ancient notions of vital life forces and "essences" that hearkened back to Paracelsus and Aristotle, which were slowly but surely seeping into the contemporary biological debates on the origins of matter. At this point in the middle of the eighteenth century, Enlightenment science still tended to favor so-called "preformationist" models of generation that regarded permutation between the species as a ridiculous fable, a superstitious remnant of medieval transformative magic like alchemy: rather, in the preformationist view, each embryo in an egg or seed contained within itself all its parts, already preformed, and in its most reductive formulations, growth was not an organic process but simply a mechanical expansion in size, according to the preformed designs of Deity. Thus Albrecht von Haller, the Swiss anatomist, could claim that the first human mother, Eve, contained two hundred million tiny human beings in her ovaries, all preformed at the moment of her creation—the future posterity of humankind simply needed to one day get bigger over generations and time. But, as Denise Gigante writes, "[S]uch a scenario included no room for unexpected change or invention. Regardless of whether what preexisted in the egg was design or an actual miniature of the animal, advocates of preformation considered generation a mechanical realization, by way of nutrition, of already articulated parts."[61]

The gradual reemergence of organic epigenesist theories in the 1740s challenged these precepts, and allowed for variation, change, and—perhaps most crucially—a way of acknowledging external environmental factors that affected embryonic development and species growth, an acknowledgment that opened an intellectual space for later work on natural selection. Swedenborg clearly, and unequivocally, supported the epigenesist position. His material on ovum and egg development in the *Principia*, where he compares the inception of the egg's matter to a formless, Ovidian chaos, already placed him in the camp of epigenesist embryologists. Later, in the theology, the organic flow of influx that causes in all species "a constant striving for the human form, a form that it gradually takes on," explicitly refutes preformation theories (which at that time, pre-Darwin, were confusingly—-for our present ears—called "evolutionary theories," even though they were in principle quite the opposite of our modern sense of that word).[62] In a wonderful moment of spiritual vision in Swedenborg's *Divine Love and Wisdom*, he is led by angels to watch the growth of an embryo inside the human womb—a sight that reads like a spiritualized and accelerated endoscopic camera—and the unfolding description makes the human clearly formed by epigenesis.[63]

It is still quite a leap from this moment to the Darwinian breakthrough of the next century, and if there is any kind of "epigenesist poetics" (to borrow Gigante's useful phrase) within Swedenborg's writings, it still remains within the providential telos of a divine plan. Nevertheless, as Robert Richards has demonstrated, the connection between Darwinian evolution and the earlier kinds of epigenesis forms a vital link in an intellectual genealogy that descends from eighteenth-century biology into the organicism of the Romantics and beyond.[64] Such a link also becomes apparent when diagramming Emerson and his relationship to Swedenborg's thought (which Emerson could have first encountered, as observed, under the banner of "theosophical influx" in his father's library). For the reprint of Emerson's seminal *Nature* in 1849, at the same time that he was immersing himself in Swedenborg's scientific work and befriending the Swedenborg translator James John Garth Wilkinson in London, Emerson drafted a new epigraph for his inaugural philosophical debut that had first appeared in 1836 (and kick-started the Transcendentalist movement). The new, six-line poem concisely epitomizes both a correspondential view of nature and the evolutionary echoes in the transmutation of forms. Emerson's poem might even, arguably, intentionally echo and rework the line from Swedenborg's *Divine Love and Wisdom* quoted above, that nature had a "constant striving for the human form, a form that it gradually takes on." As Emerson famously puts it:

> A subtle chain of countless rings
> The next unto the farthest brings;
> The eye reads omens where it goes
> And speaks all languages the rose;
> And, striving to be man, the worm
> Mounts through all the spires of form.[65]

In the next chapter, Emerson's trans-corporeal worm, mounting through these "spires of form," will be situated in the context of Swedenborg's numinous language of things, and Emerson's growing, wondering sense of that "occult relation between man and vegetable."[66]

3
Radical Correspondence
Emerson's Ray of Relation

> It needed some great poet to marry nature again to the mind for whom men who had lost the link. And this service the next ages will probably owe to Emanuel Swedenborg.
>
> —Ralph Waldo Emerson

Ralph Waldo Emerson's first contact with the thought of Swedenborg may have been in his father's library. William Emerson owned a copy of *A Theosophic Lucubration on the Nature of Influx* (1770)—an early British translation of Swedenborg's *De Commercio Animae et Corporis* (Interaction between soul and body), which had first appeared in 1769. As detailed in chapter 2, this little theological treatise set out to clarify Swedenborg's position on the relation between nature and spirit, and recapitulates earlier visionary material where Swedenborg had witnessed scientists (Sir Hans Sloane and Martin Folkes) in the afterlife debating the various positions taken on spiritual and mechanical causality in the eighteenth century: from adherents of Leibnizean "pre-established harmony," to Aristotelian partisans of Spinoza's monads, to Cartesian arguments for the doctrine of occasionalism. All three of these positions were, of course, central to Swedenborg's own characteristic panentheism, though it is telling that his angels declare heavenly favor for "spiritual influx," the side proposed by the Cartesians.[1] Swedenborg's problem in *The Nature of Influx* was inherited by Emerson's later Romantic generation, made subsequently all the more complex by advancements in natural sciences that provoked dilemmas about divine telos in the natural order, and vexed the relation between nature and spirit. When Emerson made his literary and philosophical debut with *Nature* (1836)—the little blue-covered book that announced, more or less, the dawn of American Transcendentalism—he had planned to follow the text with a volume devoted to nature's counterpart that would be entitled "Spirit": a book that, tellingly, never coalesced. Emerson abandoned (or at least, greatly modified) his early commitment to the Idealist premises of the 1836 *Nature*, and the subsequent wealth of essays and lectures he produced over the next thirty years can be seen as careful, pragmatic negotiations between his commitment to natural science on the one hand, and his deeply religious response to nature as a place for spiritual experience, on the other: "In the woods, we return to reason and faith."[2] Though Emerson has been seen as anticipating the evolutionary breakthroughs of Alfred Russel Wallace and Charles Darwin—Emerson was among Darwin's early enthusiastic American readers—he nevertheless retains his sense of nature as intrinsically ecstatic, in the fullest, religious meaning of that word.[3]

A crucial conduit connecting spirit to the messy stuff of matter, apparent in his 1836 *Nature* and remaining a consistent, life-long preoccupation for Emerson, was language. Against the post-Lockean empiricists of the early nineteenth century who continued to insist that (human) language was a completely arbitrary system of signs, randomly shaped by the vagaries of culture, Emerson voices a longing for a kind of human language intrinsically shaped by environmental forces and the raw, spiritual power of nature, experientially apprehended as the reading of a sacred script. Words were not merely accidental sounds formed by social conventions, Emerson declared, but part of the very fabric of the cosmos. "It is not only words that are emblematic," as he puts it in the "Language" chapter of *Nature*, "it is things which are emblematic. Every natural fact is a symbol of some spiritual fact. Every appearance in nature corresponds to some state of mind, and that state of mind can only be described by presenting that natural appearance as its picture."[4]

This signal moment in *Nature* betrays Emerson's deep indebtedness to Swedenborg's doctrine of correspondences and hews fairly closely, in some cases almost word for word, to a small text on correspondences by the French Swedenborgian Guillaume Caspar Lencroy Oegger.[5] Swedenborgian correspondences facilitated Emerson's dislocation of spiritual ethics away from organized religion—the institution of the Unitarian church that Emerson formally left in 1832—and into the edifice of nature. It is precisely Swedenborg's "theological determinism," in turn, his too-deep attachment to the "Christian symbol," that led to Emerson's strongest critiques of, and distancing from, his representative mystic; "Swedenborg is retrospective, nor can we divest him of his mattock and shroud."[6] While scholars of Transcendentalism, particularly those who have paid close attention to the religious dimensions of Emerson's project, have acknowledged Emerson's appropriation of Swedenborgian ideas, Swedenborg nevertheless has tended to be omitted from any serious consideration within environmental or ecocritical (re)appraisals of Emerson.[7] Sometimes this has been due to the critical tendencies noted in this book's introduction: that a secular bias has lingered in environmental historiography, and colored ecocritical readings of Transcendentalism that, following Lawrence Buell's influential *Environmental Imagination,* have pitted the Idealist, anthropocentric Emerson against the (better, more holistic) ecocentric naturalism of Henry David Thoreau. Fredrik Brøgger writes, for example, with no lack of hyperbole, that Emerson's *Nature* is the most "arch-villainous text" in a history of anthropocentric environmental hubris that characterizes American Romanticism—Brøgger finds the "Language" chapter of *Nature* to be an especially egregious illustration of the "direct ideological appropriation and cultural colonization of the environment."[8] Such blanket assumptions have created a secular supercessionist narrative about the appearance of nineteenth-century environmental imagination that both belies a more complex exchange of ideas, and has further suppressed Thoreau's more overtly religious and mystical tendencies.[9]

Swedenborg's "radical correspondence" between visible things and our human thoughts helped Emerson delineate an "occult relation between man and vegetable."[10] Even as such rhetoric was ridiculed in Emerson's own day—perhaps most infamously with Christopher Pearce Cranch's satirical sketch of Emerson as a spindly, detached "transparent eyeball," floating on the blithe New England air—this occult "ray of relation" is nevertheless consonant with some of the recent work in the new materialisms that attempts to move ontologies of matter beyond Cartesian dualism and its dead mechanism. Emerson both suggests the necessity of a new kind of posthumanist writing about nature (as called for at the close of the "History" essay, from 1841), as well as articulating a form of panpsychism: a nondualistic unity of the faculties of perception with the stuff of nature. In Emerson's later thinking, nature becomes the "incarnation of a thought," the activity of "mind precipitated." Such statements can be fruitfully engaged with more recent preoccupations with panpsychism, especially in the work of David Skrbina and Freya Matthews, whose thinking, in turn, has inflected the so-called "object-oriented ontology" of Graham Harman and others working under the banner of the new materialisms.[11] To read Emerson in light of contemporary panpsychist philosophy and its phenomenological analogues in ecocriticism is to take seriously (not ridiculously) the actual vegetables in Emerson's "occult relation" with nature; that at a deep level of symbiosis, if not biosemiotics, Emerson is not really "alone and unacknowledged. They nod to me, and I to them." Or, as Emerson puts it elsewhere in his journal, in a longer meditation prompted by Swedenborg, "spirit is matter reduced to an extreme thinness—O so thin!"

Emerson came to own and annotate at least fifteen titles of Swedenborg's science and theology (a total of twenty-one separate volumes), and his frequent reference to other works, such as the *Arcana Caelestia*, suggest a familiarity with additional volumes not present in the extant collections of Emerson's library in Concord and at Harvard's Houghton Library.[12] Emerson also subscribed for over two decades to the monthly magazine published by the Boston Swedenborgians, the *New Jerusalem Messenger*, where he closely followed—among other articles—a long multi-issue scientific review of William Kirby and William Spence's *Introduction to Entomology* (1826), written by Sampson Reed.[13] Emerson collected works about Swedenborg written by other Transcendentalists (such as Henry Hedge), and owned a further four volumes related to Swedenborg penned by Sampson Reed, at least six other volumes on Swedenborgian topics by James John Garth Wilkinson, as well as several commentaries and essay collections by Henry James Sr.

The collective heft of these thirty-two plus volumes, give or take, circulated within a conversational exchange and discursive space that was essential for the development of Emerson's ideas. Conversation, or "table talk," became constitutive, with Emerson moving the recorded snippets of speech from his journals, into the performance of his lectures that were composed out of the compiling of fragments and aphorisms taken out of his journals, and then subsequently layered into the dense, compressed prose of his published essays.[14] Conversation became, for Emerson, a kind of "meta"-topic that he reflected on and thought about in the abstract: an indice to the "thinking about thinking," or what Richard Rorty and others have called the "'idea' idea."[15] Conversation, when it was good, "is a series of intoxications" that produces (divine) inspiration.[16] Emerson's conversations occasionally attempted to include the other-than-human as interlocutors, to "hear the rats in the wall, see the lizard on the fence, the fungus under foot, the lichen on the log," and to overcome the fact that "there is no record of any word or sign that has passed" between us and them: "broader and deeper must we write our annals."[17] Emerson wished for his books—his first collection of essays that he imagined titling "Forest Essays"—to "smell of pines and resound with the hum of insects."[18] Swedenborgian correspondence theories gave Emerson a structure for an ecopoetics that saw language as fundamentally a natural, even evolutionary, force, akin to geological formations: "[L]anguage is fossil poetry. As the limestone of the continent consists of infinite masses of the shells of animalcules, so language is made up of images, or tropes, which now, in their secondary use, have long ceased to remind us of their poetic origin."[19] Correspondence was to commune and communicate with the other-than-human; conversing with nature had left traces of primal residues within language that poetry alone was capable of excavating.

Keeping in spirit with the dilating rings of Emersonian conversation, this chapter will discuss Emerson's engagement with Swedenborgian thought primarily through his relations with two notable nineteenth-century Swedenborgians, Sampson Reed (1800–1880) and James John Garth Wilkinson (1812–1899), both of whom Emerson was close to, and exchanged many ideas with. In contrast to chapters 1 and 2, which proceeded more or less chronologically through the development of Swedenborg's ideas, this kind of discursive approach will better serve to locate the vitality of "influx" and "correspondence" across the spectrum of Emerson's varied ecopoetics—as articulated in his journals, essays, lectures, and letters. By "ecopoetics" I mean to cue a self-reflexive conception of language-making that is attuned to how actual, physical environments shape words, and vice versa; a poetics in the sense of *poesis*, a making that brings something new into the world.[20] Emerson's ecopoetics evolved over time, and this chapter pays particular attention to several late essays and lectures that chart Emerson's nature-as-language tropes as they move far beyond the Idealist parameters of the early 1836 *Nature* debut, and even come to suggest a radically decentered view of the human subject. In the late Emerson, language functions as a natural system of composing and decomposing agents, like the living and dying of organisms in a coral reef. To engage with ecopoetics was to encounter a vastness that unsettled individual agency and suggested our own finite, earthbound mortality. These late considerations, percolating in and around the inchoate metaphysics of his incomplete "Natural History of the Intellect" project, reveal Emerson at his most chthonic (in Donna Haraway's sense), his most "earthy" feeling that words reveal a nature which "decomposes all her harvest for recomposition."[21] Even at this juncture in Emerson's mature thought, we find him making frequent recourse to Swedenborg, translating the Scandinavian's immanental theology into an ecopoetics of natural bodies and words that circulate, live, and die into one another.

Sampson Reed's "Words Like Things"

Swedenborg's ideas first definitively arrived, for Emerson, on a presumably hot and muggy August day in 1821 when Emerson heard Sampson Reed give a valedictory address at Harvard University. Reed was in the same graduating class as Emerson's brother, William; also likely to be on the proscenium with William were Theophilus Parsons, later a seminal conduit for Sarah Orne Jewett's Swedenborgianism, as chapter 5 takes up, as well as William Worcester, the father of the San Francisco Swedenborgian minister whose Arts and Crafts church is enmeshed in the early history of the Sierra Club. Though Emerson was but eighteen, he later recalled that it was as if "an oracle" were speaking, whose sybilic lips "showered words of fire" onto Emerson.[22] The experience was catalytic—perhaps as singularly transformative as the later moment of *aha!* insight that Emerson experienced in 1833 at the Jardin des Plantes, where the ordering of the specimens in the cabinets of natural history crystallized an insight into the compositional creative power of the mind and its relation to "the upheaving principle of life everywhere incipient in the very rock aping organized life forms."[23] As soon as the talk was over, Emerson sought the speaker out, obtained a copy of the address through his brother William, and transcribed it, word for word, into his journal. The pungency of this first encounter with Reed and his articulation of key Swedenborgian concepts like correspondence is suggested by Emerson's sharp recollection of the day fifty years later, and also in how Reed's writing weathered Emerson's falling out with Reed himself, and a mutual disenchantment between Emerson and his "Swedenborgian druggist," as he came to derisively refer to Reed. Reed had been unable to find employment as a Swedenborgian minister—the fledgling New Church organization was just getting off the ground—and Reed became a highly successful apothecary instead of a theologian. Reed and Emerson ultimately parted ways over the Fugitive Slave Act, as Emerson sadly notes in his journal, with Reed coming to support Daniel Webster's opprobrious political accommodations.[24] A late, unpublished manuscript poem by Emerson laments how his "demure apothecary" turned into a "sleek deacon of the New Jerusalem," a "blind man's blind man."[25]

What originally electrified Emerson in Reed's valedictory "Oration" was its clear articulation of an epistemology based on intuition that flagrantly challenged the Lockean models that had so desiccated Unitarian theology in the early nineteenth century. "Locke's mind will not always be the standard of metaphysics," Reed challenged at the end of his talk.[26] As intuition was a cognitive faculty everyone was capable of accessing, Reed democratized an idea of genius, having an innate potential for perception of Divine Power, open to all. Great men like Luther, Shakespeare, and Milton refract aspects of the Divine Light we all carry within us.[27]

There is no distinct reference to Swedenborg in the "Oration," although Swedenborg's correspondence theory clearly lies behind Reed's unusual pronouncements about language and nature. He calls for a new history of poetry that would make evident where "words make one with things, and language is lost in nature." There is the implicit suggestion in the "Oration," made more overt in Reed's subsequent *Growth of the Mind* (1826), that language is not just shaped by natural processes and environment, but operates as a natural environment in and of itself. "Thoughts fall to the earth with power," Reed writes elliptically, "and make a language out of nature." He continues this proto-etymological interest in the origins of words into a revealing commentary about the Genesis story of Creation:

> Adam and Eve knew no language but their garden. They had nothing to communicate by words, for they had not yet the power of concealment. The sun of the spiritual world shone bright on their hearts, and their senses were open with delight to natural objects. In the eye were the beauties of paradise; in the ear was the music of birds; in the nose was the fragrance of the freshness of nature; in the taste was the fruit of the garden; in the touch, the seal of their eternal union. What had they to say?[28]

In contrast to this picturesque, preverbal state of Eden, where gardening was a kind of language, Reed lashes out in a critique of contemporary ecclesiology that would surely have resonated with Emerson, as it anticipates the caustic sting of Emerson's own later "Divinity School Address." "The churchyard is a graveyard," Reed declares. "The bell which calls men to worship is to toll at their funerals, and the garments of the priests are of the color of the hearse and the coffin. Whether we view [the church] in the strange melancholy that sits on her face, in her mad reasonings about truth, or in the occasional convulsions that agitate her limbs, these are symptoms, not of life, but of death."[29] According to Laura Dassow Walls, Reed here "effectively gave Emerson permission to cut loose, as he did, immediately," from the scriptural Gospels as the sole source of spiritual authority.[30] The radical theological implications of Reed's talk, combined with its poetics that pushed for a new language "where words are one with things," mark the text as one of the first Romantic manifestos in American letters.[31] Years later, signaling its importance to Transcendentalism, Elizabeth Palmer Peabody chose to reprint the speech as part of her 1849 *Aesthetic Papers*, where it appeared alongside Thoreau's "Resistance to

Civil Government." Although Reed became embedded within the fledgling Swedenborgian church (much to Emerson's chagrin), the iconoclastic nature of his early "Oration" stuck in Emerson's mind, and accounts for Emerson's consistent sense that "to the withered traditional church, yielding dry catechisms, [Swedenborg] let in nature again."[32] Reed was, as Emerson later put it in a poem, "a bard of nature to these Times / With words like things."[33]

Reed's ideas in the short "Oration" were more fully developed five years later in the longer treatise entitled *Growth of the Mind*. This, too, became a text that Emerson continually revisited, and it could even be said that Reed's pithy distillation of Swedenborgian doctrines ultimately fascinated Emerson more than any particular work by Swedenborg ever did. As the title suggests, *Growth of the Mind* presents an organic theory of consciousness and language development. Instead of a tabula rasa, Reed writes, "the mind is originally a most delicate germ, whose husk is the body. . . . The process of learning is not by synthesis, or analysis. It is the most perfect illustration of both. As subjects are present to the operations of the mind, they are decomposed and reorganized in a manner peculiar to itself, not easily explained."[34] Reed reiterates his prior claim that primeval man had "a language not of words, but of things," and roots the origins of words in the physical impression left by the nonverbal image: the pictographic quality of language means that for modern man, in our fallen state, the poet plays a unique role in illustrating Divine Truth through a recovery of this original natural imagery.[35] Drawing on Swedenborg's correspondences, Reed charges the figure of the poet with a new vocation to restore our lost connection to nature, foremost by emphatically grounding this "language of things" in observed scientific fact. Echoing William Wordsworth's Preface to *Lyrical Ballads* (1803), another Romantic manifesto he was surely acquainted with, Reed proclaims poetry to be "the bulwark of science," a moral litmus test that should indicate the true state of science and religion.

These sentiments find their later analogues in Emerson, particularly the alignment of poetic language with scientific truth. Walls, for example, writes about this kind of synthesis at work in Emerson: "Nature is a hieroglyphic language, it is natural history that can provide the 'key' or 'dictionary': since nature is a language, the natural historian must be a poet as much as a scientist, even as the poet must also study natural history."[36] There are also several key environmental implications in Reed's aesthetic that we might see as contributing further to Emerson's own emerging ecopoetics. The poet's revelatory recovery of the "thingness" of words was wholly dependent on an internal revolution of spirit that altered how the poet perceived and observed phenomenal nature. Reed describes the need for an active reciprocity with the voices of nonhuman nature that, firstly, demands a stillness of the mind in order to listen (thus reformulating the charming image from his "Oration" of a silent Adam and Eve, with gardening as their lingua franca). "The very stones cry out," Reed writes, "and we would do well to listen to them."[37] In a similar manner,

> because we are unwilling to hear, we find it necessary to say so much; and we drown the voice of nature with the discordant jargon of ten thousand dialects. Let man's language be confined to the expression of that which actually belongs to his own mind; and let him respect the smallest blade which grows and permit it to speak for itself. Then may there be poetry, which may not be written perhaps, but which may be felt as part of our being. Everything which surrounds us is full of the utterance of one word, completely expressive of its nature.[38]

Reed's transformation of nature into a speaking text that "permits it to speak for itself" leads to the most innovative and surprising argument made in the whole text. Because our language should correspond with nature, and one does not find anywhere in the natural world perfect symmetry and exact repetition, but only endless variety, thus do "rhymes add nothing to poetry, but rather detract from its beauty." Reed proceeds to formulate what is probably the earliest and clearest exposition of a free-verse poetics, at least on the American continent:

> In the natural world we find nothing which answers to [rhyme and meter], or feels like them, but *a happy assemblage of living objects springing up,* not in straight lines and at fixed distance, but in God's own order, which by its apparent want of design, conveys the impression of perfect innocence and humility. . . . The poet should be free and unschackled as the eagle; whose wings, as he soars in the air, seem merely to serve the office of a helm, while he moves on simply by the agency of the will.[39]

Reed's thesis had a marked impact on Emerson's own poetic activity, pushing him to experiment (however unsuccessfully) with free verse and nonrhyming poems in his journals.[40] The further step to Walt Whitman is obvious, or at least clarifies Emerson's immediate appreciation for understanding what Whitman was trying to do with language when *Leaves of Grass* first appeared in 1855.[41]

Because organic nature determined the form of language, and nature was an ongoing process of constant animation, in Reed's schema humans perpetually needed to take account of the changing phenomena before them. Reed writes that if the mind undertakes "the habitual observation of the relationship between things," while respecting the inherent "order in the Cabinet of Creation," the mind is able to dissolve objects and their relationality in order to rearrange them into new harmonies "with all the precision of crystallization."[42] The metaphor of thinking that Reed deploys here is striking: it concretely anticipates Emerson's famous insight into the mind's organizational powers that was stimulated by the cabinets of natural history in Paris. His choice of the word "crystallization" also suggests Swedenborg's known work as a crystallographer, a specific word and process that Emerson often resorts to for describing Swedenborg and his "shooting spiculas of thought."[43] Reed foregrounds environmental perception as a dynamic process: by allowing the "the stones to speak" and subordinating human language to "respect the smallest blade which grows," Reed comes close to modern environmental phenomenologists who have also emphasized an active interplay between "the perceiving body and that which it perceives."[44]

Composting the *Maximus Homo*

Reed modeled the organic evolution of human consciousness as a fundamentally semiotic relation to an environmental surround in a way that approximates the German biologist Jakob von Uexküll's later conceptualization of an *Umwelt*.[45] Emerson harked back to Reed's formulations about mind, language, and matter, their coterminous development, long after his published critiques of Swedenborg had appeared with *Representative Men* in 1850. In 1870, Emerson was invited to offer a course of inaugural lectures at Harvard that formed the basis of their subsequent graduate school of Arts and Sciences; he chose to title his course of seventeen talks "A Natural History of the Intellect." The talks represented the culmination of a lifetime project to draft a "New Metaphysics" that would treat the patterns of human thought as a natural entity, shaped by the evolutionary forces of climate and weather, and show thinking (and language) to be sedimented and scored like fossils or geological strata, which followed the same laws of nature that left such traces; it was an attempt to bring a kind of spiritualized "deep time" to the history of thinking as much as it tried to undo the abstract, disembodied Cartesian dualism that had persisted in nineteenth-century philosophy. While, regrettably, Emerson never brought this late ambitious project to fruition as a published collection of essays—the project being composed and drafted, it should be noted, in his twilight years, with Emerson facing the inevitable fact of his senescence—the 1870 Harvard lectures nevertheless constitute an important development in Emerson's thinking: a materialized kind of Idealism, far different from the static Platonism that permeates his 1836 *Nature* debut. Furthermore, in spite of the lecture papers for the Harvard course not having survived, earlier drafts of related lectures, as well as highly detailed notes from attendees of the lecture course, give ample evidence of Emerson's last major efforts to "annul that adulterous divorce which the superstition of many ages has effected between the intellect and holiness," as he declaimed in "The Method of Nature" (1841).[46] In these remnants and records, Emerson occasionally comes back to Sampson Reed's *Umwelt*-like ideas, and here and there references Swedenborg in his copious quotations and citations. The beginning of the first Harvard lecture apparently included a paraphrase of one of Reed's aphorisms from *Growth of the Mind:* "[I]t is the development of mind which makes the science of mind."[47]

This development of mind was organic, even "vegetable." As Emerson puts it in another lecture related to the "Natural History of the Intellect" project: "All things grow; in a living mind, the thoughts live and grow; and what happens in the vegetable happens to them."[48] This echoes some of the contemporary inflections of panpsychism, particularly with those critics, such as Matthews and Skrbina, who have paid careful attention to the spiritual valences of panpsychism.[49] For Sampson Reed, as Gillian Osborne astutely observes, "memory and all that is bound up with it, has a resolutely physical component, as regular and organic as the digestion of food or the circulation of blood. As much as possible, Reed seems interested in thinking of the mind as matter."[50] If mind was matter, then the reverse was also true: matter contained mind, even sentience. As Emerson expanded and built upon this trope, it sometimes dilated out to a surprising kind of animism in the earth, a "vibrant matter" we more often typically associate with Thoreau at Walden.[51] As Emerson continues in the same inaugural lecture:

> As thus, it is easy to take vegetation as the type of power; *to represent the world as a plant,* and the particles plants,—and the vegetable function may easily be traced through all parts of nature, and even in the functions of mind, where Freedom seems to suspend the brute organic action, *so it would be easy, with the ancient mythologists, and early theorists, to find the world an animal, which repeats in colossal the economy of animation,* which has locomotion, perspiration, inspiring and expiring of air and water, assimilates food, and draws to it with intelligence all that suits its constitution.[52]

"If the earth is seen as animate in itself," asks Skrbina, "what consistent ontology could refrain from accepting full panpsychism?"[53]

Not surprisingly, as Emerson entertained the notion of the earth as a giant, encompassing living entity (also informed by his encounters with Alexander von Humboldt's concepts of the *Kosmos*), his related essays and lectures from the "Natural History of the Intellect" trajectory show a revival of an interest in Swedenborg's literary conceit of the *Maximus Homo*, or "universal human," which was discussed at length in chapter 2.[54] Swedenborg's vision of the entire spiritual cosmos was not only as a giant creaturely figure, à la the Adam Kadmon of Kabbalah, but also as a discursive community: "Swedenborg," Emerson writes in "Quotation and Originality" (1868), "threw a formidable theory into the world, that every soul existed in a society of souls, from which all its thoughts passed into it, as the blood of a mother circulates in her unborn child."[55] Chapter 2 placed this signal moment in Emerson, where our participation in language becomes an unsettling force that depersonalizes individual subjectivity, into dialogue with the figure of the BwO—or "Body-without-Organs"—from the poststructural philosophy of Gilles Deleuze and Félix Guattari. Furthermore, in "Quotation and Originality," Emerson figures the production of language as an organic activity of "suction" akin to the prolific movement of "flies, aphides, gnats, and innumerable parasites"—with the inevitable implication that, following Darwin, language will ultimately decay and decompose, profligately, and accordingly produce new variations of life forms. This cycle of living and dying, blindly sucking, is akin to the random generation of the millions of cells that compose a coral reef (a metaphor for language-making that Emerson harks back to in this essay and elsewhere): "language is a city, to the building of which every human being brought a stone; yet he is no more to be credited with the grand result than the acaleph which adds a cell to the coral reef which is the basis of the continent."[56] "Quotation and Originality" turns on a deft play between material decomposition-as-linguistic-recomposition, the turning of others' old dead words (the language we are born into, what Stanley Cavell has called the "mother-tongue" in his work on the Transcendentalists) into new forms: "Nature decomposes all her harvest for recomposition," as the last sentence in Emerson's essay reads.[57] In "Quotation and Originality," Emerson figuratively "composts" the body of Swedenborg's *Maximus Homo* into a dynamic image of our collective human participation in language as a natural(ized), earthy force. To speak or to write is to encounter the earthbound conditions of our own mortality. Sampson Reed may have specifically helped stimulate this chthonic dimension of Emerson's late ecopoetics. Osborne points out how Reed's theory of the mind was strikingly different from other early nineteenth-century Romantic treatises, due to its Swedenborgian epistemology that demolished the self and its agency as a kind of illusion. "Unlike the versions of genius traditionally associated with Romanticism," she writes, "Reed's genius did not enlarge or glorify the individual, but enabled diminishment, and eventually, decomposition, a becoming compost for God."[58] Reed articulates this process as a kind of holistic symbiosis, of words "getting swallowed up in the source from which they came . . . so that language is lost in nature."[59]

Influx, Instinct, and the Garments of Thought

Reed was not the only contemporaneous Swedenborgian to strongly impress Emerson at the outset of his career. In the 1830s Emerson studiously read Guillaume Caspar Lencroy Oegger's *Le Vrai Messie* (1829) and *Essai d'un Dictionnaire de la Langue de la Nature* (1831). The colorful Oegger, a Catholic priest who had been confessor to the French queen, had been sent by the Catholic Church in the 1820s to spy on what were held to be dangerously radical mesmerist and spiritualist circles that various Swedenborgians in Paris had been hosting. Instead of reporting back his findings, Oegger became convinced that Swedenborg was, to the contrary, a genuine seer, and after undergoing a secondary kind of conversion experience himself in London, he felt he had been called by the Messiah "to reacquaint the world with the language of nature to which Swedenborg's doctrine of correspondence provided a key."[60] Oegger subsequently abandoned the Catholic Church, and went on to become a controversial proselytizer of Swedenborg's doctrines. Emerson avidly read Oegger's *Le Vrai Messie* in 1833 in a manuscript translation that Elizabeth Peabody had undertaken, and subsequently copied extensive sections of that manuscript into his journal.[61] As with Reed (and it seems that Reed and Oegger wrote independently of one another, and were not aware of each other's coterminous Romantic investigations), Oegger used Swedenborg's writings on correspondence to conclude that nature was the original source of all language, the "quarry" from which words were originally hewn, in his memorable phrasing.[62]

If Reed could be said to have provoked Emerson's "vegetization" of language and mind, Oegger offered a more complete account of Swedenborgian correspondences as a kind of hieroglyphics, whereby words melted into the ideogrammatic, material images of things. In an incredible moment in Emerson's journal—one that espouses an almost Surrealist valuation of the spontaneous and irrational—Emerson jots down reflections on a stimulating conversation about theology that he had held with his brother, which betrays his indebtedness to Oegger's symbolic theories:

> Charles wonders that I don't become sick at the stomach over my poor journal yet is obdurate habit callous even to contempt. I must scribble on if it were only to say in confirmation of Oegger that I believe I never take a step in thought when engaged in conversation without having some material symbol of my

> proposition figuring itself incipiently at the same time. My sentence ends in babble from a vain effort to represent that picture in words. . . . While I thus talked I saw some crude symbols of the thought with the mind's eye, as it were, a mass of grass or weeds in a stream of which the spears or blades shot out from the mass in every which direction but were then curved round to float all in one direction. When presently the conversation changed . . . then saw I as soon as I spoke the old pail in the Summer Street kitchen with potatoes swimming in it, some at the top, some in the midst, some at bottom, and I spoiled my fine thought by saying books take their place according to specific gravity 'as surely as potatoes in a tub.' I suppose that any man who will watch his intellectual process will find an image, a material image, contemporaneous with his every thought and furnishing the garment of his thought.[63]

Emerson's choice phrasing, his search for a material image to "furnish the garment of thought," is a direct paraphrase of an idea in Swedenborg's *Theosophic Lucubration on Influx* (which, perhaps, we can imagine a young Emerson reading in his father's library).[64] The text contains numerous metaphors to illustrate the dynamic relationship between the spiritual and the natural, including an exposition as to how "ſpirit clotheſ itſelf with matter; and, indeed, as a garment is to a man, ſo is the organical body to his ſoul, which he may be properly ſaid to put on, to wear it, and to put it off at the time of death."[65] This was Swedenborg's so-called "Doctrine of Affections Clothed" that Emerson makes recourse to in his journals throughout the 1830s.[66]

Swedenborg's influx theory became a key part of Emerson's conceptual vocabulary about nature. It is present in the "furnishing the vestment of thought" from *Nature,* but one also finds it repeated in the later essays that tack away from Emerson's earlier Idealism. The word "influx" had first entered Emerson's journals in 1836, specifically in context of his (re)reading of Sampson Reed.[67] Emerson subsequently made the concept central to his definition of Transcendentalism: "[T]he Transcendentalist adopts the whole connection of spiritual doctrine. He believes in miracle, in the perpetual openness of the human mind to new influx of light and power; he believes in inspiration, in ecstasy."[68] Here, Emerson seems to have absorbed the way that Swedenborg's influx doctrine not only entailed a panentheistic flow of vitalist power, from inner to outer, that animated all of nature—"all nature is the rapid efflux of goodness executing and organizing itself," as Emerson puts in "Circles"—but also suggested new modes of cognition, a "new influx of light and power" in the human brain. The more the mind naturally opens itself to acknowledging, a priori, the reality of a spiritual world that maps out and shapes perceptions of phenomenal nature (and here, Swedenborg is close to Immanuel Kant's descriptions of the operations between phenomena and noumena), the greater is the capacity for perceiving analogies and correspondences in the organic world, and synthesizing discrete facts into truths that form larger wholes.

This is less a process of rational comprehension than it is a faculty of intuition and feeling, as Emerson's later work in and around the "Natural History of the Intellect" project makes clear. The best humans, Swedenborg had written, who come to occupy the highest heavens are those who the moment they hear truth intend to do it and live it—"knowing instantly, from an influx from the Lord, whether the truth they are hearing is actually true."[69] As Anders Hallengren puts it, paraphrasing a section of Swedenborg's *Theosophic Lucubration*, "a spiritual influx, which exists quite apart from the understanding, i.e. from thought, like that which makes animals act appropriately, is called instinct. That influx directly determines [human] actions."[70] This analogy between influx and instinct can be found throughout Swedenborg, and draws by coextension an implicit line between the human and the other-than-human. Emerson had closely followed a long, multipart review essay that Sampson Reed had written about William Kirby and William Spence's influential *Introduction to Etymology*, which appeared over a three-year period in the *New Jerusalem Messenger*, between 1828 and 1830. Emerson's subsequent references to this important work on insects often hark back to Reed's Swedenborgianized review. In the review essay, Reed sets forth influx as a kind of intuitive, spiritualized instinct: "When the human will shall be brought into coincidence with the Divine Will, then will also reason live from instinct and look towards it. . . . Our instincts will no longer pass under the name of animal propensities by way of expressing their degraded inferiority to reason."[71] Hallengren stipulates that this is probably "the background to Emerson's reference to instincts in connection with self-reliance, and the reason why [Emerson] suddenly linked Swedenborg to Goethe and Carlyle when his major notion of individuality was developing."[72]

For the purposes of unfolding Emerson's ecopoetics, "influx" can be read as a dynamic mode of knowing that creates a complex and intimate kinship between human and the other-than-human. As influx is an emanating, creative flow (ultimately of Divine Love) that is everywhere at the same time, the more regenerate the human perceiver, the nearer she comes, instinctively, toward intuiting that this source behind herself and all appearances binds them together into relationships as part of a larger interconnected whole. This moment of perception is very different from the Platonic view of natural objects as mere shadows of fixed higher essences, and might be seen as approximating Maurice Merleau-Ponty's late attempts to expand his phenomenology of the body to include "the flesh of the world."[73] For Emerson, Swedenborg's influx that spiritualized instinct and privileged intuition bore an important ramification for the perception of nature: namely, that the "NOT ME" beyond the self—to paraphrase Emerson paraphrasing of Samuel Taylor Coleridge and Johann Gottlieb Fichte at the opening of *Nature*—was not best comprehended from a scientific perspective (although this, too, was indispensable), but was rooted in a *feeling* aligned with the intuition of the artist, and with aesthetic experience in general.[74]

Emerson's ready gravitation toward Swedenborgian influx took place within a broader stream of his attention to water and flow. Starting, perhaps, with his early lecture on "Water" (1834), the aqueous nature of shimmering, sliding surfaces and streams quickly became a central organizing metaphor for his entire philosophy. "We live amid surfaces," he pragmatically advises in "Experience" (1844), "and the true art of life is to skate well on them."[75] In Branka Arsić's brilliant reading of Emerson's so-called "ocean ontology" of fluctuations, mutations, and waveforms, she reads Emerson's obsession with water as ultimately leading to a radical dissolution of the self. Where other critics of Emerson—such as Stanley Cavell and Sharon Cameron—have called this the "impersonal" at work in Emerson, with its negative "assemblage of forces and powers (of perceptions and sensations, emotions and thoughts) that work within the personal," Arsić rather argues "in favor of a personal identity contrived of tensions, divides, and gaps that make it permanently 'tremble.'"[76] As Swedenborgian influx not only coursed through the vagaries of the human mind, but was a constitutive flow to all of nature, this trembling of a self's assemblages—its "tensions, divides, and gaps"—wobbled open to include the other-than-human in its currents, its departures and arrivals.

With the whole cosmos adrift and mutating, nothing is fixed or stable. Emerson's "ocean ontology" meant he did not see science as a metaphysical substitute for the loss of religion—the natural sciences provided no surety of an Absolute. The close of Emerson's "History" essay is remarkable for its provisionality in this regard, and the great leeriness it expresses toward extending language's representative powers over the other-than-human—a quite striking shift away from the purported "villainous" anthropocentrism of the earlier *Nature*. Emerson's direct invocations of influx in "History" might be read vis-à-vis Merleau-Ponty's "flesh" as a tacit acknowledgment of perceptual reciprocity in nature, and a movement on the part of Emerson toward a more biocentric perspective. It further crystallizes Joan Richardson's observation that Emerson demonstrates how "truth happens to an idea in a reciprocal relationship with an environment."[77] As Emerson writes at the essay's finale:

> I hold our actual knowledge very cheap. Hear the rats in the wall, see the lizard on the fence, the fungus under foot, the lichen on the log. What do I know sympathetically, morally, of either of these worlds of life? As old as the Caucasian man,—perhaps older,—these creatures have kept their counsel beside him, and there is no record of any word or sign that has passed from one to the other. . . . Broader and deeper we must write our annals,—from an ethical reformation, *from an influx of the ever new*, ever sanative conscience,—if we would trulier express our central and wide-related nature, instead of this old chronology of selfishness and pride to which we have too long lent our eyes. Already that day exists for us, shines in on us at unawares, *but the path of science and of letters is not the way into nature*. The idiot, the Indian, the child and unschooled farmer's boy stand nearer to the light by which nature is to be read, than the dissector or the antiquary.[78]

This "influx of the ever new" into the human mind effects a moral regeneration of perception that opens Emersonian rays of relation "to those neighboring systems of being" in the homely creatures of the lizard and rats in the walls. Emerson calls here for a new kind of (natural) history writing—"broader and deeper must we write our annals." If we are to read Emerson's "Poet" as anticipating Whitman in direct and specific ways, the close of "History" calls forth Thoreau at Walden Pond.

Wilkinson's Primitive, Derivative Words

Thoreau came into Emerson's circle precisely as Emerson's criticisms of Swedenborg and the Swedenborgians were becoming sharper, distancing Emerson from his initial enthusiastic embrace. He had written to Carlyle in 1834 how the Swedenborgian New Church "was deeply interesting, as a sect which I think must contribute more than all other sects to the new faith which must arise out of all."[79] Yet, a few years later, all Swedenborgians had sectarian "lockjaw," and Emerson is irritated at the "immense arrogancy and subtle bigotry" of Sampson Reed's little church. Swedenborg was killed by ideas "as moth to a flame," his writings were "a dull prismatic blur," he was "like a bear who fattens in the dark & in the cold."[80] Sometimes Emerson's critiques of Swedenborg's theology coincided with his growing interest in Eastern religious literature (an interest that also parallels the emerging intensity of Emerson's friendship with Thoreau in the 1840s).[81]

Swedenborg's correspondence theory, in fact, has been read by scholars of Transcendentalism as a decisive factor in the subsequent rift that developed between Emerson and Thoreau. According to Joel Porte, correspondential theory was a "pious fraud" that Thoreau inherited from Emerson, which had to be perforce exorcised, a process most apparent in the writing process behind *A Week on the Concord and Merrimack Rivers* (1849). This book shows Thoreau (according to Porte's influential reading) consciously eschewing Emersonian Idealism to embrace "the reality of things" (in Thoreau's words). And yet, this dialectical reading that pairs a materialist Thoreau, allied with the truths of science, against an idealizing Emerson out of touch with the real, elides how Thoreau's view of language and nature was a consistent attempt "to read the spiritual meaning of the physical world," in Brian Harding's words.[82] This inflection, Harding claims, had deep roots in Swedenborgian theology and conceptualizations of language. For Thoreau, Swedenborg was not "the airy mystic" but "a man whose vision transformed the commonplace."[83] Harding's recovery of Swedenborg for Thoreau's nature aesthetics relies on a careful attention to Thoreau's responses to James John Garth Wilkinson, the London-based homeopathic doctor, spiritualist-poet, publisher of William Blake, and translator of Swedenborg.[84]

Wilkinson's writings on Swedenborg, and especially his English (co)translations of Swedenborg's scientific works, played an essential role in the Transcendentalist reception of Swedenborg in the 1840s and 1850s. Indeed, without Wilkinson's illumination of Swedenborg's science, it would be easy to imagine Emerson growing so dissatisfied with Swedenborg's "tinsel and gingerbread" theology that Swedenborg could have even fallen out of the "Representative Men" pantheon. A closer attention to Wilkinson's mediation of Swedenborg for Emerson (and also simultaneously to others in his circle, like Thoreau) elucidates how Swedenborg came to be viewed as, in Emerson's words, "a leader in that revolution, which, by giving to science an idea, has given to an aimless accumulation of experiments, guidance and form, and a beating heart."[85]

Wilkinson's friendship with Emerson began in 1846, when he appears to have sent Emerson a copy of his English translation of Swedenborg's *Animal Kingdom*.[86] An earlier date of communication is very likely, however, as Wilkinson and Henry James Sr. had begun their robust transatlantic correspondence at least two years prior. Wilkinson had published an important article in the *New Jerusalem Messenger* transcribing Coleridge's annotations to

Swedenborg's scientific works in 1841, an event that could not have failed to attract Emerson's keenest interest. Given the closeness of Emerson to the James family, and his subscription to and regular readings of the *New Jerusalem Messenger,* Emerson probably became interested in Wilkinson at least by 1843.[87]

Wilkinson's biographical accounts of Swedenborg, both his 1841 essay for the *Penny Encyclopedia* and his longer full-length book published in 1849, were used by Emerson as sources for his lectures on Swedenborg and the essay that ultimately appeared in *Representative Men.*[88] Wilkinson's 1849 biography was one of the first to postulate a coherent connection between Swedenborg's scientific projects and his subsequent theological career, braiding the two interests together.[89] "Another synthesis effected by Swedenborg," as Wilkinson writes in the biography, "is that of poetry with reason and science." The subsequent passages elaborate on the necessary role the imagination performs for science, for uniting disparate parts together into larger wholes—a classic function of the Romantic imagination, we might note, that Karl Kroeber claims is a crucial antecedent for ecosystemic holism.[90] In a passage reminiscent of the end of Emerson's "History," Wilkinson calls for the imagination in this "full stretch and lustiest life" to "touch the facts the living God has made." Wilkinson continues: "If you want to understand a beetle, look at it with all the imagination through the glass of the universe; translate it into a mineral, into a vegetable, and into a man; run it along its own line of genera and species, and let it catch illumination from them all; and when you have enlarged it from this associated empire, its atomic theory will be palpable and distinct: every habit, limb and entrail will be a self-evident proposition."[91] Wilkinson's valuation of this synthetic power of the scientific imagination is what leads Emerson to claim in "Representative Men" that Wilkinson has a "coequal vigor of understanding and imagination comparable only to Lord Bacon's."[92]

Wilkinson's particular commentaries and expositions on Swedenborgian correspondence are also noteworthy in that regard, as they emphasize the material, natural conditions necessary for its analogical structure. In Wilkinson's essay on "Correspondence" that appeared in Peabody's *Aesthetic Papers,* for example, he stresses how a genuine correspondence does not place substantial materiality in deference to a higher ideal realm, but rather unites the natural and spiritual together into a single moment of intuitive perception. As Harding puts it, "far from subordinating the physical to the spiritual, the theory of correspondence was seen by Wilkinson as a means of making them one—of producing 'conjunction.'"[93] Accounting for why Emerson would later write that Wilkinson was a reincarnated Francis Bacon, Wilkinson here in the *Aesthetic Papers* calls correspondence a "nexus of creation" that "manifests what Lord

Bacon calls transitive instances" when "creation is taking place."[94] Wilkinson's interpretation of correspondences as Baconian "transitive instances" makes them play out a (proto)pragmatist sense of truth.

Reading Wilkinson's influential articulation of correspondence theory as pragmatist further discloses the importance of lived experience that underwrites Wilkinson's definition of how correspondential analogy operates, as it also explains Wilkinson's sharp critique of Idealist Continental philosophy as misanthropic, a point with which Emerson seems to have resonated. Wilkinson writes in "Correspondence,"

> There is, however, one caution which cannot be too often enforced in the prosecution of analogies and correspondences. It is, that both terms of the intellectual equation must lie within some sphere of experience, or no conclusion will be valid from one to the other. . . . Experience, therefore, is indispensable in both spheres; and, if there were no actual experience of the spiritual world, there could be no safe conclusion, except a negative one, from the natural world to the spiritual. Therefore correspondence does not engender, but simply follows experience.[95]

The stress on lived experience here reencapsulates a longer argument Wilkinson had made via Swedenborg in his "Introductory Remarks" to Swedenborg's *Outlines on the Infinite* (a segmented part of the larger *Principia* project) in 1847. Emerson highly valued this particular essay by Wilkinson; he gushed to Margaret Fuller that that "you ought to see [Wilkinson's essay]. Tis much the best piece of criticism on Modern Opinions, especially Transcendentalism, which has yet been written."[96] This was almost surely what Emerson had in mind when he later wrote in *Representative Men* that Wilkinson's various prefaces "throw all contemporary philosophy of England into shade"—a further oblique (but very public) jibe, it would seem, at Carlyle, whom Emerson had moved away from as Carlyle became ever more cantankerous and conservative. (Carlyle, interestingly, privately held Wilkinson at a somewhat skeptical distance.)[97] Wilkinson's "Introductory Remarks" to the *Outlines* is a robust critique of Kantian Idealism premised on Swedenborgian theology, arguing that transcendental philosophy, in a scrambling attempt to counteract skepticism and materialism, completely neglected human and natural facts, creating a version of the mind that was ultimately "disembodied life," as Wilkinson puts it, and therefore unintelligible. Kant's insuperable divide between noumena and phenomena led to a philosophical Gordian knot where "both spiritual and natural experience are shaken to the base, and all scientific vision of the deeper parts of nature, is set down as a dream, in which, the figments of the mind are mistaken for the unattainable truth of 'things in themselves.'"[98] Wilkinson's proposed solution to this Kantian dilemma is a necessary encounter with (and accounting for) the "natural foundation" of truths in "our mechanical age," that shows the interdependence of both spiritual and natural realms. The domain of thought is naturalized; ideas are a force just as tangible and real an impression on the mind as the five senses.[99]

Wilkinson's attempt to reconcile an Idealist position with the material "outward universe" can be mapped in conjunction with Emerson's own shift away from the Idealist positions originally taken in *Nature*, where the natural world is wholly subject to the higher moral needs of man, to a more complex and simultaneously pragmatic sort of ecopoetics incipient in the later essays collected in *The Conduct of Life* (1860), and also in the lectures and unfinished material for the "Natural History of the Intellect" project. Emerson had learned his lesson from Darwin, among others: "Nature is no sentimentalist. . . . [Nature] does not cosset or pamper us. We must see that the world is rough and surly, and will not mind drowning a man or a woman; but swallows your ship like a grain of dust. . . . The habit of snake and spider, the snap of the tiger and other leapers and bloody jumpers, the crackle of the bones of his prey in the coil of the anaconda—these are in the system, and our habits are like theirs."[100] In the wake of all this bloody data provided by natural history, Emerson asks, pragmatically, "[H]ow shall I live?" "Our geometry cannot span these extreme points, and reconcile them. What to do?" Emerson's turn toward experience owes a debt here to Wilkinson, whose deliberate weaving of correspondence into a natural, organic base informed by natural history became part of Emerson's later conceptualizations of nature. The Pragmatist strand could be devolved back even further to Sampson Reed's gnomic aphorisms in *Growth of the Mind* that "truth, all truth, is practical," that "truth is the way in which we should act," and how truth "adheres to objects."[101] Reed's source for such statements lies in Swedenborg's elaborations on a doctrine of "use" and truth: Swedenborg allowed Emerson and others to "spiritualize use" away from the strictly human (and even explicitly commercial, free-market) valences that utilitarian philosophy had acquired in the early nineteenth century, a shift essential for the subsequent emergence of an ethically oriented Pragmatist philosophy in the later work of Charles Saunders Peirce and William James.

Finally, to circle back to Thoreau: Emerson was by no means the only Transcendentalist to respond to Wilkinson and his Swedenborgian ideas in ways we might read as either proto-pragmatist or proto-ecological. In 1851, a year after supposedly breaking free from the "pious fraud" of Swedenborgian correspondence doctrines (at least according to Joel Porte), Thoreau's journal details a sustained and telling reading of Wilkinson's *The Human Body and Its Connexion with Man, Illustrated by the Principal Organs* (1851), a book-length study—perhaps Wilkinson's best monograph—on the universal correspondence of the human figure as a kind of master signifier for scientific fields. Thoreau meditates at length in his journal, in response to Wilkinson's words:

> Wilkinson's book to some extent realizes what I have dreamed of a return to the primitive analogical & derivative senses of words—His ability to trace analogies often leads him to a truer word than more remarkable writers have found.— . . . The man of science discovers no world for the mind of man with all its faculties to inhabit—Wilkinson finds a *home* for the imagination—& it is no longer cast out and homeless. All perception of truth is the detection of an analogy.—we reason from our hands to our head.[102]

Swedenborg's correspondential vision, his hieroglyphics of nature, diffused here through Wilkinson's vivid prose, stimulate Thoreau to dream "of a return to the analogical and derivative sense of words." This dream becomes the foundation for *Walden* (a book, and a place, that is both a "port" and a "foundation"), its repeated etymologies that trace language through a series of puns that turn back to linguistic roots, when words were once "one with things," as Sampson Reed phrased it. This climaxes in *Walden* in its penultimate "Spring" chapter, where Thoreau's meditation on the melting, muddy banks of snow push him to a profoundly evolutionary perspective on language, as many have taken note. According to Philip Gura, "while Thoreau's notion of the relationship between language and nature can hardly be termed 'transcendental' in the Emersonian or 'correspondential' in the Swedenborgian sense, he arrived at his philosophical and imaginative conception of language by working through the linguistic premises of such theories."[103] This acknowledgment should correct the common ecocritical contention that Thoreau's hermeneutics of nature primarily stemmed from his encounter with texts by Coleridge. Along with Gura, I would argue that these overlooked readings in Wilkinson, Reed, and Swedenborg were just as significant for revolutionizing Thoreau's approach toward experiencing nature as "living poetry," a "written revelation," as the jubilant end of *Walden* puts it.

Thoreau exemplifies what Emerson had made implicitly possible through his rendering of Swedenborg as a "representative man" for America at its mid-century point, a "lens through which we read our own minds." Swedenborg's ideas were to be used, to be pragmatized: "[T]he power to which [representative men] communicate is not theirs. . . . Great men exist that there may be greater men. The destiny of organized nature is amelioration." Positioned by Emerson as a figure who straddled the divide between nature and language, Swedenborg became integrated into the later environmental poetics of American Transcendentalism. Amplified by Emerson's circle, Swedenborgian "influx" and "correspondence" were broadcast as components of aesthetic experience that went on to gravitate and coalesce around the emergence of an environmental imagination. The groundswell of an American environmental movement in the preservation and conservation efforts at the century's close was energized by John Muir's vision of aesthetic value in wild spaces that inculcated religious, ecstatic wonder. This has at its core a concept of the transcendental self engaged in a hermeneutical decoding of nature, suddenly released by an epiphanic perception of correspondence and felt connection with a larger whole. To Muir, his beloved Sierra Nevada displayed "divine hieroglyphics"; they were holy places where one could witness nature "eternally flowing from use to use, beauty to yet higher beauty; and we soon cease to lament waste and death, and rather rejoice and exult in the imperishable, unspendable wealth of the universe."[104] That an eighteenth-century Scandinavian mystic ended up coloring Muir's influential proclamations about the American West is one of the stranger and more interesting stories in the evolution of American ideas, revealing the lubricity of how theological typology could slip out of institutional religious categories and into the experience of wilderness. That the numinous character of the wild remains so hotly contested by environmentalists and ecocritics further suggests how the ray of relation that Emerson originally drew from Shakespeare to Swedenborg continues as a live wire into our present, echoing a Romantic aesthetic legacy that refuses to relinquish its grip on the ways we continue to imagine ourselves and the world around us. The political efficacy of this ecopoetics of the sublime and its many problematics—the "trouble with wilderness," as William Cronon terms it—is taken up in the next chapter, which addresses Muir and the circle of Californian Swedenborgians who formed a critical, generative context for the early conservation work of the Sierra Club.

4
Heralds of a New Gospel
John Muir and the San Francisco Swedenborgians

> I know something of what went on in Muir. For him, quite simply, the spirits of the wild were angels, who bore him on their wings through perilous places.
>
> —Mary Austin

Ralph Waldo Emerson's call for a new kind of nature writing at the end of his essay "History" can be read not only as adumbrating the later appearance of Henry David Thoreau's *Walden,* but also as catalyzing agent for a young John Muir, who began studying Emerson's writings in the early 1860s, and who would later become, by the century's close, the iconic grandfather of the modern American environmental movement. Muir's formative role in establishing the Sierra Club in 1892 and his related political advocacy for the protection of wilderness areas throughout the American West heralded the beginning of an environmental imaginary that gave shape to the establishment of the National Park Service, in 1916, and even later on, the Environmental Protection Agency (in 1970).[1] Muir's consistent description of landscape as a sacred text, descried both by science and an ecstatic, felt apprehension of its beauty, is unthinkable without his formative encounter with the doyen of American Transcendentalism.

Muir famously met Emerson in 1871, when the young naturalist was working in Yosemite Valley. In spite of Muir's enthusiastic overtures, the elderly

Emerson would not leave the comfort of his entourage for a night of camping in the rough with Muir in a sequoia grove. They nevertheless spoke of many things, including Emerson's telling Muir about the mystic Emanuel Swedenborg and his reputed conversations with the dead.[2] Their meeting prompted Muir to start a substantial rereading of Emerson, returning to the essays that had so electrified him when he first encountered them in the library of his mentor Jeanne Carr in Wisconsin. Some time after their Yosemite encounter, Muir wrote to Emerson reminding him to send along a certain Swedenborgian text that Emerson had promised. Emerson replied with a warm, enthusiastic letter that enclosed a small volume—none other than Sampson Reed's *Observations on the Growth of the Mind.*[3] Emerson had just finished the "Natural History of the Intellect" lecture series, which indicate a revived interest in Reed, his "Swedenborgian druggist" (as chapter 3 took up); the still-unknown Muir, like Reed before him, would have borne out Emerson's contention in the lecture series that original genius percolated in the margins, among the obscure. Muir enthused to Jeanne Carr that Reed's little book was "full of the fountain of truth."[4]

This may not have been Muir's first substantial encounter with Swedenborgian thought, however. He might have encountered earlier an artistically inclined Swedenborgian named Joseph Worcester who had moved to California (from the East Coast) originally for health purposes, to be in the region's salutary climate. Worcester had spent intermittent periods in Yosemite Valley from the 1860s onward—even working, for several months in 1864, for James Mason Hutchings, who would later become Muir's employer.[5] The charismatic Worcester was soon to become a major gravitational force, forming the center of an important coterie of influential painters, poets, architects, and writers in a burgeoning arts scene in the San Francisco Bay area. This group of "Worcesterites," as they were sometimes called, were the progenitors of the West Coast Arts and Crafts movement, a Ruskinian attempt to "bring nature indoors" through architecture, furniture, and the fine arts.[6] The therapeutic dimensions of the Arts and Crafts aesthetic philosophy—its back-to-nature healing modalities that manifested in architectural innovations like the sleeping porch—shared in the same formative milieu of wilderness recreation that germinated John Muir's Sierra Club. Worcester's establishment of a new building for the San Francisco Swedenborgians in 1895 was infused with Swedenborg's doctrines of correspondence and influx; Muir is reported to have attended, on rare occasion, and some of the church parishioners were involved in the early days of the Sierra Club (William Keith, the painter, for example, was a charter member).[7]

Scholarship on John Muir has not paid significant attention to the books by Swedenborg and Swedenborgian texts like Sampson Reed's *Observations on the Growth of the Mind* that Muir collected, read, and annotated over several decades. Muir's distinctive marginalia—neat pencil-line jabs, and in some cases, exquisite filigree sketches and doodles of alpine landscapes—can be found in three extant Swedenborg(ian) texts now housed in the Muir archive at the University of the Pacific, and further references to Swedenborg occasionally pepper Muir's unpublished corpus. Overall, Swedenborg could be said to have reinforced Muir's scriptural tendencies to translate nature into a kind of bible, a numinous wilderness text that, if apprehended with the right reverence and wonder, would unfold a revelatory message. Due, in part, to the persisting influence of the ecological critiques of Christianity made by Lynn White and others, environmental historians have sometimes attempted to locate an Eastern strand in Muir's religious inclinations, attributing his spiritual view of nature to Buddhist, Taoist, or even Confucian origins.[8] As Dennis Williams observes, this has obscured how "Muir's preservationist ideology emerged as a natural outflow of his mystical Christianity."[9]

If we amplify the aesthetic contexts that swirled around Worcester, Keith, and the San Francisco Swedenborgian Church, which developed in tandem with the political advocacy of the Sierra Club, we can better delineate the theological dimensions of three salient aspects of Muir's writing. First, Muir understood the destruction of nature as a kind theological desecration, a blasphemy that Muir responded to, in turn, by drawing on the well-trod ground of Christian polemics to shape his public conservation advocacy. Secondly, perhaps surprisingly, Muir's making of the Sierra Nevada into modern bibles broadened his capacity to acknowledge the other-than-human, decentering the anthropocentrism that he had inherited from Romantic literary traditions (Emerson, William Wordsworth) that were seminal for his ethics and aesthetics. Under the charm of immanental influx and correspondence, Muir expanded his sense of personhood: "[A]ll God's people, however serious and savage, great or small, like to play. Whales and elephants, dancing, humming gnats, and invisibly small mischievous microbes—all are warm with divine radium."[10] The human presence was simply one of many in a larger divine semiotic of nature speaking, evolving, radiating life. And yet, thirdly, while Muir's own variant on the "language of things" he encountered in Sampson Reed may have fomented a more biotic kind of environmental imagination, the spiritualized construction of wilderness also rendered it as a blank page, and thereby participated in a problematic erasure of indigenous peoples already present throughout the California Sierras. Muir's writing is constitutive of the "trouble with wilderness" famously anatomized by William Cronon; what has been sometimes lost in the ensuing wilderness debates is the theological habitus behind Muir's "reading" of places as pages of sacred Scripture.[11]

This chapter will accordingly proceed in three distinct sections; instead of moving chronologically, and untangling in a linear fashion Muir's different engagements with Swedenborg and the San Francisco Swedenborgians, each section will focus on a particular body of materials—paintings, buildings, and a text—where Muir's inchoate environmental imagination intersected with the work of Keith, Worcester, and others. In the first, William Keith's landscape paintings will be juxtaposed with Muir's rhetoric to consider how Keith's aesthetic functioned as a foil for Muir's development as an environmental writer and activist. Keith's paintings were critically important for Muir's first debut as a public speaker; they formed the literal backdrop for the Sierra Club in the 1890s when the group would informally meet in Keith's studio in San Francisco, and, in spite of a falling-out and mutual disaffection as Keith's Swedenborg-influenced canvases became increasingly diffuse, abstract, and atmospheric, Muir still enlisted Keith's paintings in the struggle to preserve the Hetch Hetchy Valley at the very end of Muir's life: Keith's paintings, thus, can be seen as visual bookends to Muir's entire career.

The second section turns to the San Francisco Swedenborgian Church, where Keith installed important landscape paintings as permanent wall murals (completed in 1895, three years after the official founding of the Sierra Club). While the church has duly received attention from art and architectural historians for its significance in the development of a distinctive Arts and Crafts style in California, rarely has the space been situated in relation to the work of Muir and the Sierra Club. As a veritable temple to nature—or more precisely, a temple dedicated to the spiritual *within* nature—the building probably also refracts the mutual influence of Muir on Swedenborgian imaginaries. Many of the building's distinct, unusual features—the rough-hewn tree branches that retain their bark and act as the supporting pillars for the roof—have their ample corollaries in Swedenborgian theology, as we will see, but they also echo the vernacular architecture of Yosemite Valley, including Muir's cabin that famously had a stream running through it, which Worcester may have seen in person. A related structure, a beautiful house that Worcester designed for himself that anticipates some of the natural elements of the later church, will also be briefly considered, as the domestic space generated both important paintings by Keith and became the later home of the writer Jack London, who lovingly called it "the Bungalow." This house of Worcester's in the East Bay hills, in fact, was where London penned his Naturalist classic *The Call of the Wild* (1903). These

Swedenborgian spaces—with their sleeping-porch verandas and rough natural materials—inculcated a therapeutic regime whereby wilderness aesthetics were seen as a healing tonic for the debilitating effects of modernity. Even within Jack London's most secularized and Darwinian aesthetic, the spiritual valence of wilderness allowed a residue of an original religious impulse to remain. The Swedenborgian church and Worcester's cottage where *The Call of the Wild* was born, should be seen as part of an aesthetic constellation that helped birth literary Naturalism and an American nature-writing tradition out of the Romantic Transcendentalism of Emerson into the work of Muir, and beyond. Acknowledging these contexts should complicate, at the least, the typical literary historiography that sees American Naturalism as evolving out of the secularizing drift of encounters with post-Darwinian science.[12]

London's highly popular *Call of the Wild* was but one of many new kinds of writing about animals that burgeoned at the turn of the century as humans were reconsidering their place in the cosmos and the natural order of things. Muir's own famous tale of a dog, "The Adventure of a Dog and a Glacier" (1897), which was later expanded into *Stickeen* (1908), was based on Muir's spectacular adventure with a special dog on a stormy day in Alaska. He claimed, curiously, it was "the hardest thing [he] ever wrote."[13] The final section of this chapter pays heed to Muir's draft notes for *Stickeen,* where he jotted down how his favorite dog was "like Swedenborg a Herald of a New Gospel." When we turn to the narrative of *Stickeen,* however, it is not clear just what kind of good news is being evangelized. How is Stickeen the silent dog like a Swedenborg, the Scandinavian mystic? The ambiguity in the story might be one place to locate Muir's own ambivalence toward indigenous peoples; Stickeen was so named because of the dog's connection to the Stickeen tribes of the Pacific Northwest. I argue for reading the dog's odd vanishing at the end of Muir's narrative metonymically, as the literary figuration of a caesura in Muir's enchantment of wilderness that depended on an erasure of native peoples from the landscape. The inscription of a sacred language of things into a sublime natural space needed to be prefigured as a blank page, devoid of human presence. In the end, the Americanization of Swedenborg's original doctrine of correspondences into a language of things could produce beautiful, proto-environmental architecture and painting for the Bay Area Swedenborgians, but it also facilitated an appropriation of land into landscape that was buttressed by an ideology of Manifest Destiny, and the complex representation of Yosemite as the reclamation of a prelapsarian, premodern sort of garden: this was to view Yosemite through the lens of the Puritan poetics of John Milton, as Mark Stoll's work has so brilliantly excavated.[14] Keith's paintings and Muir's writings are entangled in the aesthetics of what W. J. T. Mitchell has called the "the dream work of imperialism," that is, an acculturated depiction of a natural landscape oversaturated with nationalist fantasies and their narratives.[15] These embodiments of America as "Nature's nation," as Perry Miller put it, depended on a visual logic mutually constructed by text and image, converging around a theological episteme of nature-as-Scripture, as the next section takes up.[16]

Vague Jumbles of Paint

The artist William Keith first met Muir in Yosemite in 1872, introduced by Muir's longtime mentor Jeanne Carr, who had been much impressed by Keith's paintings. Keith had first heard of Muir from Ralph Waldo Emerson, whom Keith had befriended during a long sojourn in New England (Keith's first wife, Elizabeth Emerson, was a distant relative of Ralph Waldo).[17] Not surprisingly, Keith and Muir became fast friends: both were Scottish immigrants, with parents who had been steeped in a kind of harsh Calvinist Protestantism (Presbyterianism) that their sons would later eschew, and both "shared the view that nature should be protected from human destruction, that creatures large and small had moral worth, that beauty in nature was a divine gift, that wilderness had spiritual value," as Ronald Limbaugh puts it.[18] The friendship commenced thirty years of collaboration, commentary, and critique of one another's work. In spite of their later differences—caused, in part, by the growing influence of Swedenborgian ideas on Keith's mature, post-epic painting—Keith remained Muir's "poet-painter" whose depictions of California's Sierra Nevada were a "kind of bible of the mountains," as Muir favorably put it.[19]

Keith's painting frames the commencement of Muir's vocation as a public speaker on behalf of the environment, when Muir was thirty-eight years old, in 1876. Muir had had no formal training in public lecturing and was extremely nervous at the invitation to speak to the Literary Institute of Sacramento (the talk was to be on the process of glaciation in the Sierras). A large canvas by Keith of California mountains was set up near the podium for Muir to gaze upon while he talked, and the effect of the landscape was such that Muir "completely forgot himself and his audience, only remembering that he was to make clear some wondrous mysteries."[20] The talk began Muir's bona fide career as a lecturer on behalf of the environment and a public champion of the earth sciences, and the Keith painting accompanied Muir's speaking engagements on several other occasions.[21] This anecdote suggests how Keith's landscapes functioned for Muir as a veritable window onto nature, mimetically reproducing the healing spiritual power of wilderness; as such, they provide a kind of mirror for refracting the power of the visual in Muir's own writing, where thick layers of ekphrastic prose attempt to re-create the dazzling optic dimensions of Muir's wilderness experience. Both Muir and Keith continued a form of the "panoptic sublime" that could be said to have formalized in American nature aesthetics with both Emerson's 1836 *Nature* and Thomas Cole's famous painting *The Oxbow*, created the same year.[22] Even more so than Emerson's, Muir's influential writings operate along optical registers, unspooling long chains of visual association related to light and color: "glittering," "gleaming," "azure," "purpled haze," "white, glowing, irised falls and cascades," for example. Often the act of seeing land-as-landscape through the lenses of the sublime and the beautiful becomes a spiritualized moment of ecological connection, as in this well-known epiphany from Muir's *My First Summer in the Sierra:*

> No Sierra landscape that I have *seen* holds anything truly dead or dull, or any trace of what in manufactories is called rubbish or waste; everything is perfectly clean and pure and full of divine lessons. This quick, inevitable interest attaching to everything seems marvelous until the hand of God becomes visible; then it seems reasonable that what interests Him may well interest us. When we try to pick out anything by itself, we find it hitched to everything else in the universe. One fancies a heart like our own must be beating in every crystal and cell, and we feel like stopping to speak to the plants and animals as friendly fellow mountaineers.[23]

The famous line here about "when we try to pick out anything by itself," ironically, has often been plucked out of context to underscore Muir's ecological sensibility (it makes a good pithy aphorism, nice for bumper stickers and coffee mugs), yet it comes nested within Muir at his most theological, where he presses the land to become both a visual landscape and a semiotic, speaking text—a page "perfectly clean and pure and full of divine lessons." This syntactic fissure between text and image is significant; the gap creates a characteristic tension that is part of the efficacy of Muir's prose. Muir continues Sampson Reed's Swedenborgian project to recover a language—both verbal and visual—whereby "words were one with things."

Four years prior to his public lecture debut, as Muir read and annotated Reed's *Observations*, when he encountered the specific section where Reed expounds on a wish for "a language not of words, but of things" Muir responded by penciling in the margins a minuscule landscape depicting an alpine slope with pine trees, pitching down to a mirrorlike lake (fig. 2). Precisely at the juncture where Reed wants words to exceed and move on to some kind of representative space beyond language—into "things"—Muir responds with the

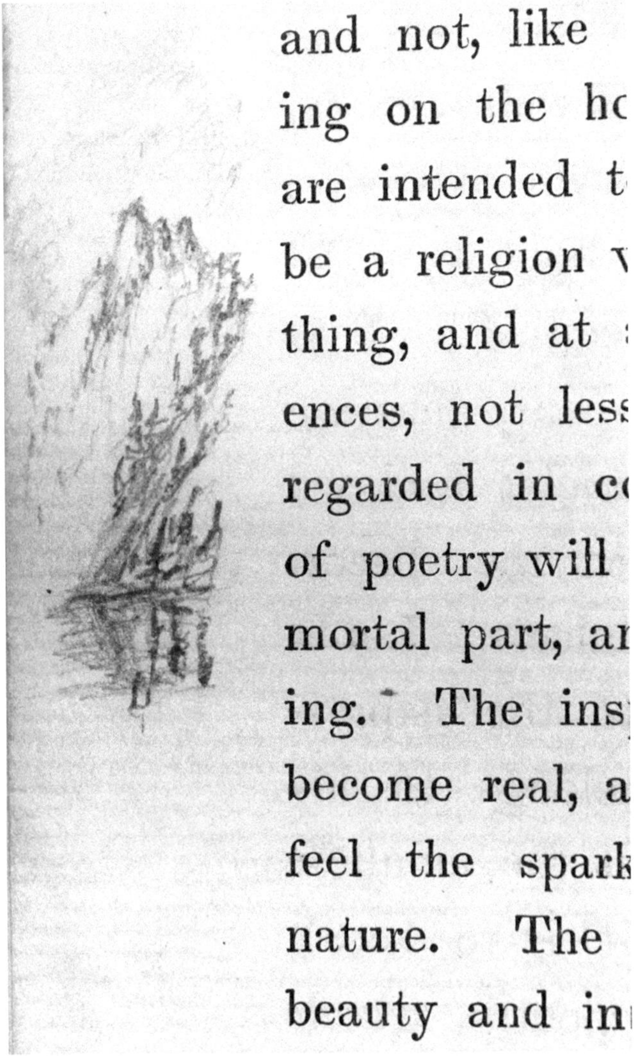

and not, like
ing on the h
are intended t
be a religion
thing, and at
ences, not less
regarded in c
of poetry will
mortal part, a
ing. The ins
become real, a
feel the spark
nature. The
beauty and in

FIGURE 2. John Muir's marginalia to Sampson Reed's *Observations on the Growth of the Mind*. (Chicago: Meyers & Chandler, 1867; John Muir Papers, Holt-Atherton Special Collections, University of the Pacific Library. © 1984 Muir-Hanna Trust. Photograph courtesy of Michael Wurtz, 2019)

visual. Yet, it is the visual frame of the landscape in Muir's prose itself, the rhetoric of the sublime and the beautiful noted in the prior passage, that, in turn, provide the mechanism for a spiritual inscription, the transformation of mountains (back) into bibles and divine manuscripts. Yosemite is "a grand page of mountain manuscript I would gladly give my life to be able to read," as Muir puts it, somewhat hyperbolically.[24] This semiotic chain is lubricious and synesthetic, sliding between seeing and reading, and listening: a word becomes a seen thing, which is then made into the divine Word that speaks iconically: *My First Summer* frequently tropes on landscape as "hieroglyphic." The ekphrastic dimensions of Muir's nature writing makes it more modern, even Modernistic, than its purple-drenched private ecstasy might otherwise suggest—an emotive style that often rather recalls the Romantic pantheism of earlier authors like William Wordsworth, who is as central as Emerson was for shaping Muir's sense of nature.

The same year Muir had read Reed's *Observations*, perhaps still under the tingling influence of Emerson, Muir published his first essay in the *New York Tribune*. "Yosemite Glaciers" lays out Muir's innovative theories of glaciation in the high Sierras, and while there is little of the religious ecstasy so characteristic of his later nature writing, this early essay strikingly opens with an extended metaphor of the mountains as a book:

> Two years ago, when picking flowers in the mountains back of Yosemite Valley, I found a book. It was blotted and storm-beaten; all of its outer pages were mealy and crumbly, the paper seeming to dissolve like the snow beneath which it had been buried; but many of the inner pages were well preserved, and though all were more or less stained and torn, whole chapters were easily readable. In just this condition is the great open book of Yosemite glaciers to-day; its granite pages have been torn and blurred by the same storms that wasted the castaway book. The grand central chapters of the Hoffman, the Tenaya, and Nevada glaciers are stained and corroded by the frosts and rains, yet, nevertheless, they contain scarce an unreadable page; but the outer chapters of the Pohono, and the Illilouette, and the Yosemite Creek, and the Ribbon, and Cascade glaciers, are all dimmed and eaten away on the bottom, though the tops of their pages have not been so long exposed, and still proclaim in splendid characters the glorious actions of their departed ice.[25]

The unnamed book in this anecdote (and it is tempting to imagine it as a bible, weather-beaten by climate) deliciously slides back into a kind of materiality that renders it almost illegible: a language of things dissolving into pure thingness—"all of its outer pages were mealy and crumbly, the paper seeming to dissolve like the snow."[26] If Reed's Swedenborgian "language of things" was extended to its fullest logical consequence, going back into a time when "words were one with things," semiosis would ultimately fail, the signified and signifier collapsing into one another. Muir's metaphor-rich prose, his sustained attention to the visual dimensions of representation, ensured that a fissure remained between words and things, a *différance* (in Jacques Derrida's sense) between seeing and speaking (or writing).[27]

Muir's attunement to the possibilities of word and image within his landscape prose can be usefully situated against the evolution of Keith's painterly style. While Muir purportedly vastly preferred Keith's earlier epic canvases that meticulously reproduced details of geography, flora, and fauna—much as in the "Humboldtian" manner of Frederic Edwin Church—there are telling parallels between Keith's later subjective, even proto-Modernistic paintings that foreground a process of perception and Muir's own nature writing that takes increasing issue with the mimetic reproduction of merely the material, or "scientific," aspects of nature.[28] Most art historians agree that the death of Keith's first wife, Elizabeth Emerson, pushed a grieving Keith more into the orbit of Joseph Worcester's Swedenborgian influence, perhaps due to the Swedish seer's insistence on the verifiable continuity of life after death.[29] Worcester's ensuing impact on Keith's aesthetics was profound, as richly illustrated by Keith's surviving correspondence to Worcester that begins during Keith's long sojourn in Europe between 1883 and 1885.[30] Likely solely through Worcester, Keith came to own seventeen works by or about Swedenborg (including the multivolume English translation of Swedenborg's magnum opus, the *Arcana Caelestia*).[31] The aesthetic ramifications of Swedenborg's doctrines were further reinforced by Keith's intense friendship with the Swedenborgian landscape painter George Inness (whose paintings are discussed in more depth in chapter 5). Keith was well aware of Inness's aesthetic theories, derived, in part, from Swedenborg, long before they eventually met face-to-face—Keith's 1888 lectures on art at the University of California, Berkeley, for example, drew heavily on Inness's earlier essay, "A Painter on Painting," that had appeared in *Harper's Magazine*.[32] The two months that Inness spent in San Francisco in 1891 galvanized Keith, with Inness painting up a storm side-by-side with Keith in his studio (and Keith regularly taking Inness to Worcester's Swedenborgian church group, then meeting at Druid's Hall in downtown San Francisco on Sundays). "I no longer want to die since Inness has arrived," Keith reported to Worcester.[33]

Under all this Swedenborgian influx, Keith's painting drifted away from large-scale, detailed landscapes into more subjective and "moody"—in the words of his contemporaries—atmospheric depictions of nature, closer in spirit to the Tonalism of Inness. These later paintings tend to become smaller in scale and size, and the titles (following Inness) occasionally shift from the specificity of topical places into more poetic sorts of associations: *Golden Skies, Mood, Gray Day, Ducks in Water, Dazzling Clouds* (though Keith certainly continued to paint landscapes designating particular locales).[34] Keith no longer wanted to reproduce a mimetic fidelity to nature, but rather the record of an inner, spiritual apprehension of a phenomenological epiphany that a particular place had provided. "What a landscape painter wants to render," he wrote for a lecture in 1895, "is not the natural landscape, but *the state of feeling* which the landscape produces in himself. Under the impulse of feeling he has produced a piece of work and the feeling will have fused the material into a whole. Art is not the slave of nature."[35]

Art critics—including Muir—felt Keith's shift in style was a betrayal of his earlier commitments to a honed realism. "It is amazing that a man who knows so much, who has done so much," wrote a scathing reviewer of a gallery show in the *San Francisco Evening Post*, "should at this time of life fall into the silly error of supposing that there is more poetry in *a vague jumble of paint* than in the facts of nature."[36] Yet today, Keith's "vague jumble of paint" is seen as cannily prescient of how landscape painting helped generate later movements into abstraction and expressionism.[37] Akin to Wassily Kandinsky's synesthetic interest in how abstract painting could re-create a feeling of symphonic sound, Keith, in this late period, collected various gongs and bells from Buddhist temples, and would ring them while he painted his landscapes—often, by this point, painting a composite landscape completely from different memories, with no preparatory sketches: he wanted his colors to capture the feeling of vibratory sound, and landscapes from this period were accordingly known as Keith's "gong pictures."[38] Color was, Keith wrote, merely "a thing of vibration."[39]

Muir, apparently, would have none of it. "Why in the deuce don't you imitate nature?" he is reported to have scolded during one disappointing studio visit: "You'll never paint a decent picture until you can do that."[40] Keith, in turn, disparaged Muir in his public lectures: "Mr. Muir thinks of me as one of the lost, a son of perdition. I don't know why I drag poor Mr. Muir in so much, except that as I told you at the beginning, this was to be a personal experience, but I use him in another way: which is that a man may be a poet and a man of education and may never get beyond the outside husk of art, the imitative side."[41]

But this mutual, jocular sparring that juxtaposed their representational practices belied a deeper point of convergence. Whereas Keith's later work became more and more modern, Muir's writing, in turn, became increasingly Transcendentalist, in the Emersonian sense. As Donald Worster observes, as Muir aged, he "sounded more and more like a typical theist or Transcendentalist seeking beyond nature a God in heaven, a Creator of the world's material forms, or a great Spirit hovering over the earth."[42] Arguably, however, this kind of immanental Transcendentalism—or "vibrant" panentheism, to return to some of the conceptual terminology that was used in chapter 2—was always latent in Muir, demonstrable in his enthusiastic early annotations to Swedenborgian texts like Reed's *Observations*. In Muir's copy of James John Garth Wilkinson's mid-century biography of Swedenborg—the text with its Humboldtian "planetary picture" discussed in chapter 1—Muir underlined in the margins a typical passage where Wilkinson pauses to criticize those "hard-headed scientific men" who show "a deaf ear, and a callous heart" to the "dear (spiritual) facts of common men and women."[43] Muir's reservations about the limits of empiricism stemmed from deeply personal experience; he had himself undergone three acutely paranormal incidents that involved prescient dreams or uncanny intuitions of deaths in the family.[44] Not surprisingly, many of the annotations in Muir's Swedenborg volumes draw attention to the supernatural dimensions of Swedenborg's biography or theology that lay out an elaborate epistemology for how spirits communicate with humans: a Swedenborgian context, again, that seems to have been first suggested to Muir by Emerson. This material probably underscored Muir's own conviction that the spiritual was not another, otherworldly affair, but resolutely materialized and present in nature, embedded in the here-and-now. Like Keith's "vague jumbles of paint" that attempted to create an access to the spiritual—a state of subjective "feeling," in Keith's words—that lay incipient within the landscape, Muir's later writing often explicitly aims for creating a similar rhetorical effect on the reader. In the Yellowstone essay from *Our National Parks* (1901), which had first appeared as a serialized essay in the *Atlantic Monthly* three years earlier, Muir writes an extended passage on the activity of Nature as that of the artist, "ever working toward beauty higher and higher." After a lush and long description of landscape, Muir continues, sounding like Swedenborg or Sampson Reed as he keys a discourse of the "spiritual world": "Where may the mind find more stimulating, quickening pasturage? A thousand Yellowstone wonders are calling, 'Look up and down and round about you!' And a multitude of still, small voices may be heard directing you to look through all this transient, shifting show of things called 'substantial' into the truly substantial, spiritual world whose forms flesh and wood, rock and water, air and sunshine, only veil and conceal, and to learn that here is heaven and the dwelling-place of the angels."[45]

But how are we to read such idealizing, somewhat cloying lines in view of the Anthropocene, facing an anthropogenic future that might obliterate our very "flesh and wood, rock and water, air and sunshine?" In this book's introduction, Walter Benjamin's angel of history—his ekphrastic gloss on a watercolor by Paul Klee—was invoked as a figure of anthropogenic catastrophe (or at least, this is how it has been influentially reread by Bruno Latour). Might we imaginatively reconfigure Muir's wilderness angels here as Klee's "Angelus Novus," swept up in the catastrophe of progress?[46] Rather than a Neoplatonic evasion of matter, perhaps Muir's "still small voices" can reinstantiate the apocalyptics of Benjamin's *Jetztzeit* that breaks into history, unfolding a different messianic temporality that causes the "homogenous empty time" of modernity to implode on itself.[47] In this reading, Muir's injunction to "look up and down around you" is to see a future void where humans have vanished from Yellowstone, perhaps akin to the famous depeopled scene of ruins overtaken by nature in Thomas Cole's *Desolation* (1833–36) that closes Cole's *Course of Empire* painting cycle: a truly posthumanist, post-apocalyptic nineteenth-century American landscape. Such an inflection of Muir's text might seem less of a stretch if we recall Swedenborg's distinct articulation of heaven (or hell) as not future eschatological places, but internal states of mind in the here and now. Swedenborg's proto-Buddhism (as first anatomized by D. T. Suzuki) was grounded in the immanent materiality of nature. "Here is heaven," Muir writes, perhaps not at all metaphorically, "the dwelling place of angels." But "here" also just as easily slips into a hell populated by the capitalist devils of development and progress; "the mind is its own place, and in itself / Can make a Heav'n of Hell, a Hell of Heav'n," an equivocation that Muir (and Swedenborg) both knew from their Milton.[48]

The later stylistic parallels between Keith and Muir culminated in their last significant collaboration when Muir enlisted Keith and his paintings in the great conservation struggle to prevent the Hetch Hetchy Valley in Yosemite from being dammed. Their first joint work together had been with Keith's contributions to the massive *Picturesque California* project (1888–90), a multivolume series edited and overseen by Muir that cemented the Far West in the minds of Americans as premiere wilderness tourist destination. Now, Muir utilized Keith's paintings once again for what would become his final conservation battle, a legal fight that went all the way to Congress and shaped a kind of civic imaginary of the environment that continues to inform legal discourses around wilderness conservation and preservation. As Rebecca Solnit notes, the Hetch Hetchy battle was the first occasion whereby American citizens took a collective, national stance against unrestrained "growth, progress, and development," and the Sierra Club was transformed—veritably overnight—from being a quant local hiking club into America's first major conservation organization.[49] The failure to stop the dam would haunt the Sierra Club and other national environmental organizations for decades to come.[50] As Hetch Hetchy legally lay within the federal boundaries established by the designation of Yosemite as a national park in 1890, it would ultimately take a federal act of Congress—the Raker Act of 1913—to allow the thirsty city of San Francisco to dam the valley for a new source of water. In the years leading up to the Raker Act, Muir and the Sierra Club conducted a strident campaign to sway public opinion about the valley, and to bolster their efforts, Muir and Keith traveled to Hetch Hetchy in 1907 and again in 1909, trips which produced writing and images that were deployed in the campaign to stop the dam project. Both men were in the twilight of their prolific lives; the 1909 excursion was to be Keith's last extended stay in the high Sierras before his death in 1911, and the Hetch Hetchy paintings are Keith's final images of the mountains that had given shape to his whole spiritualized oeuvre.[51] These last paintings are not anything like the earlier epic panoramas that Muir had hailed as a "bible" of California mountains, but are typical of the late "mystical" Keith that Muir had publicly contemned: suggestive brushwork, almost gestural, and a close proximity of perspective in the frame that is counterbalanced by the blurring of light and shadow (fig. 3). In most of the Hetch Hetchy paintings from these trips, Keith has chosen to focus on foreshortened views of the cascades and waterfalls that tumble into the valley, feeding into the Tuolumne River—imagery that perhaps indirectly offers a wry comment on the utilitarian arguments over water made by the pro-damming advocates. It is hard not to find something slightly menacing, certainly foreboding, in the way the canvas draws our attention to the dynamic movement of the rushing water; the flows of color here stand in remarkable contrast to the relatively static and placid depictions of Hetch Hetchy that Albert Bierstadt had popularized earlier, in 1872. In Bierstadt's large paintings, Hetch Hetchy unfolds as a golden, edenic valley in a typically panoptic manner, with a clearly stable, single point of perspective that masters and surveys the land, embodying what Angela Miller and others have called the imperial "scopic power" that constitutes so much nineteenth-century American landscape painting.[52] Part of the stability recurring in Bierstadt's Hetch Hetchy imagery also comes from him using the more familiar topography of neighboring Yosemite Valley, which is wider and broader—Bierstadt notoriously repackaged the famous valleys peaks and rock faces as the "newer," lesser-known Hetch Hetchy, ostensibly for purposes of increasing the salability of his canvases.[53] By the time of his final Hetch Hetchy paintings, Keith had come to wholly reject the panoramic mode

FIGURE 3. William Keith, *Hetch Hetchy Side Canyon II*, ca. 1908. Oil on canvas, 22 x 28 inches (55.9 x 71.1 cm). Presented to the City and County of San Francisco by Gordon Blanding, 1941.5. (The Fine Arts Museums of San Francisco; photograph courtesy of the Fine Arts Museums of San Francisco)

made famous by Bierstadt and others from the "Rocky Mountain School," as Keith would have known them. "A panorama is not fine art," Keith declared, "and if deception were to be the test [of fine art] what could be more deceptive than some panorama. It is the way they are lighted that makes them so effective, and their bigness."[54] Keith's late Hetch Hetchy paintings accordingly never feature something like the panoptic sublime—the visual focus is rather in the rush of white water, a sense of destabilization furthered by the cropped bottom of the canvas that questions the locus of the observer in the visual field. Do we stand on land, or on water, or some indeterminate space between the two? The perspectival ambivalence captures the ambiguous future of the valley that was being so hotly contested in 1909.

Keith would have read in Swedenborg—particularly in the *Arcana Caelestia*—how floods of water corresponded to the "inundations" of falsities, of the "desolations and temptations" excited by evil spirits. Thus, the biblical flood of Noah, according to Swedenborg, was not an actual terrestrial event, but a story symbolic of a spiritual struggle for the truth within human consciousness.[55] While Keith's work resists didactic allegories or one-to-one correlations to Swedenborg, his late-phase work often captions his gorgeous sunsets and twilights with evocations of moods, or as Keith might have put it, "spiritual states"—these include, to add onto some of the titles already referenced, *Glory of the Heavens, Symphony of Peace, Revelation.* By contrast, the specificity of place in the Hetch Hetchy series was critical for their political valence as witnesses to the beauty of the valley that was in danger of being destroyed. Yet their aesthetics might also suggest a kind of transposition of private Swedenborgian mood into a public critique of the flood of "falsity" peddled by the prodevelopment lobby that saw Hetch Hetchy simply in terms of utilitarian extraction, as the easiest means for transforming San Francisco into a self-avowed "imperial city."[56] Keith's paintings, Muir's writings, contend the lie that flooding Hetch Hetchy was necessary for fueling San Francisco's rapacious growth.

Keith's Hetch Hetchy paintings were exhibited locally in San Francisco but also in 1913 on the floor of the Senate during the congressional debates around the Raker Act.[57] The public circulation of Keith's paintings must be understood as participating in a broader public campaign of Sierra Club encyclicals, editorials, and petitions to halt the dam. Muir's rhetoric became more stridently theological, transposing Calvinist discourses of blasphemy and desecration to the wilderness pages of California landscape. The prodevelopment arguments "are curiously like those of the devil, devised for the destruction of the first garden," he railed. Muir continued, working up to a sermonizing pitch at the end of his famous 1912 essay: "These temple destroyers, devotees of ravaging commercialism, seem to have a perfect contempt for Nature, and, instead of lifting their eyes to the God of the mountains, lift them to the Almighty Dollar. Dam Hetch Hetchy! As well dam for water-tanks the people's cathedrals and churches, for no holier temple has ever been consecrated by the heart of man."[58]

Already by this point, the advocacy work of the Sierra Club was aligning itself with the reproducibility of landscape photography, a relationship that would culminate in the iconic work of Herbert Gleason, Ansel Adams, and Eliot Porter in the decades to come.[59] This link between conservation and landscape photography had already been firmly cemented by the essential work of Carlton Watkins, whose stunning panoramas of Yosemite in 1861 directly inspired Abraham Lincoln to sign federal legislation protecting those lands for "public use, resort, and recreation" in perpetuity.[60] Watkins was deeply enmeshed in the literary and philosophical coteries that had originally brought Emerson to Yosemite Valley in 1871. As Tyler Green speculates, Emerson may have even cognized how the newly minted national park was a direct consequence of Emerson's own formative ideas about nature, which had so marked the thought of the Unitarian minister Starr King who, in turn, shaped Frederic Law Olmsted's writings about Yosemite that colored Lincoln's pioneering federal legislation. Emerson returned to Concord from his California trip with two stunning Watkins photographs of California mountains—Mount Shasta and the Cathedral Spires in Yosemite Valley—which he purchased from Watkins's gallery in San Francisco. The large photographs still hang in the sitting and dining room at the Emerson house in Concord.[61]

In Muir's later polemic to protect Hetch Hetchy, his essay gestures to the "photographic illustrations" that originally accompanied the 1912 text as veritable proof that the valley, Eden-like, comprised a "high-lying natural landscape garden." And yet, in terms of tenor and theological trope, Muir's heated rhetoric shares more with the spiritual implications of Keith's subjective canvases—much more, anyway, than the static black-and-white photographs in the volume that echo the panoramic stability of Bierstadt's paintings of the valley.[62] The displaying of Keith's art in the Capitol building during the Raker hearings demonstrates how a Swedenborgian aesthetic could become constitutive to American environmental imaginaries; the failure of these paintings, and Muir's words, to effectively negotiate legal regimes further point to the incommensurate ways early modern landscape in the United States was construed as both secular and sacred.[63] However, in both Keith's paintings and the contemporary restoration ecology efforts to bring Hetch Hetchy "back" to some mythic, premodern original state, the indigenous claims on the land remain ignored or absent, and the valley is (re)imagined as sacrosanct: showing the ideological persistence of our ongoing "trouble with wilderness" and some of its significant ethical blind spots.[64]

The Garden Church and Tree-Friends

Upon entering the San Francisco Swedenborgian Church, one is immediately struck by the resolute presence of trees and wood, from the large old-growth madrone trunks with their bark left on that support the roof, made of bare Oregon pine, to the assemblages of driftwood and branches that were left, like offerings, in the austere sanctuary. Though Muir seems to have made no direct statements anywhere about the structure itself where his friend Keith regularly worshipped, the church's unorthodox use of rugged trees and found natural objects parallels Muir's rhetorical transformation of wilderness not only into "divine manuscripts," but also into ecclesiastical architecture (Hetch Hetchy as a cathedral). The church modeled what would become a hallmark of the West Coast Arts and Crafts style: a bringing of nature indoors that was meant to "harmonize" (a key, recurring word for the movement) between human and natural elements.[65] As one had to, unusually, move through a large garden in order to enter the door into the sanctuary after coming through the entrance from the street, the church acquired early on the moniker as "the Garden Church" of San Francisco—an aspect of its design that may refract Worcester's and others' awareness of Japanese temple layouts, where sacred shrines are entered only by first passing through a garden with a symbolic program.[66]

The activities, aesthetics, and liturgical practices of the San Francisco Swedenborgians are embedded in a broader milieu of Californian spiritual experimentation and what Catherine Albanese and others have identified as a sort of distinct "nature religion" that began to be broadly practiced in the nineteenth century.[67] Like the many para-religious groups that functioned in extra-ecclesial spaces—such as the Sierra Club and, farther north, the Seattle-based Mountaineers—this milieu was avowedly countercultural, pitching their various back-to-nature programs as antidotes to modern industrialization, with its enervating urban alienation and ennui.[68] Muir's texts and the church space can be seen as mutually imbricated, each reinforcing a nature–culture dialectic through the careful construction of wilderness that, paradoxically, depends on the very space of culture for its fetishistic removal and therapeutic sanctity. If the garden church brings a precious piece of wilderness into the "imperial city" of San Francisco, we might find the inverse to be true of Muir's most raw and primal wilderness texts: that they rhetorically pivot toward a place of bourgeois pleasure and urban playground, indicating how intrinsic landscape tourism was to the conceptual construction of wilderness as a space apart. Muir's sensuous descriptions of the untrammeled wild were a kind of sales pitch for legal conservation and the inroads (often, literally) this enabled for middle-class leisure tourism, which could collectively enjoy and rejuvenate in a prelapsarian, precivilized Eden. Wilderness recreation was, as Evan Berry's important work reminds us, fully soteriological: a re-creation of American selves in fully Judeo-Christian frameworks.[69] But as Joni Mitchell would later lament, "They paved paradise, and put up a parking lot."

To juxtapose the church with Muir's nature writing can serve to underscore several tendencies that were typical of Bay Area fin-de-siècle nature-religion, illuminating both the efficacy of a wilderness aesthetic that facilitated a turn toward the other-than-human as well as its ethical incoherence. Some of these rich ambiguities can be immediately located in the Swedenborgian church's extensive use of trees, and of a parallel strand in Muir's writing that celebrates those humans who were, in his words, "tree-friends."[70]

Trees hold a special correspondence in Swedenborgian theology, signifying "perceptions and firsthand knowledge of what is good and true, which yield intelligence and wisdom"—indicating, thus, Swedenborg's Enlightenment project of a revelation based on science and truth.[71] In his scattered remarks on the spiritual history of humankind, Swedenborg claimed—like many other eighteenth-century neoclassicists—that premodern "ancient" peoples had worshipped the Divine outdoors in groves of trees; this was, according to Swedenborg, because of their "science of correspondences," their knowledge of the spiritual semiotics of nature that modernity had subsequently lost.[72] In view of the church's use of stained glass, gemlike in the darkened wooden interior, Joseph Worcester, the primary designer of the space, would have also been familiar with the fantastic passages in Swedenborg's *Earthlike Bodies Called Planets* that had described an interstellar visit to an alien planet where sacred architecture made out of living tree boughs twined around prisms of orange and blue light.[73]

Historians concur that it is largely Worcester's singular vision that shaped the space and guided the aesthetic decisions around its construction.[74] In addition to all the Swedenborgian materials on groves and trees, Worcester would have been aware of Sampson Reed's *Observations on the Growth of the Mind* where, as noted in prior chapters, Reed's "language of things" articulates a longing for a poetics in accord with the asymmetry and variety of the organic natural world (Worcester's father, the Reverend Thomas Worcester, had established the Swedenborgian Church in Boston alongside Sampson Reed—both had been in the same graduating Harvard class that Emerson's older brother, William, had also been part of). These tenets had not only a consequential analogue in free-verse poetics; we can locate them in the tree-lined sanctuary of the Swedenborgian church, where the entire space deliberately disrupts neoclassic parallelism.[75] This is most noticeable in the enormous off-kilter fireplace on the back wall, deliberately askew from the central axis of the nave (and what kind of church places a massive fireplace in its sanctuary?). Next to the fireplace, and adorning the sanctuary around the altar on the opposite end, the early parishioners of the church interspersed boughs and branches of different trees found in the hills and in the driftwood detritus of nearby beaches.[76] There is nothing in the chancel that would indicate the space is a Christian church: where a crucifix or other traditional iconographic referent would be placed, one finds instead a profusion of branches and leaves that appear even more stylized and flattened against the concrete stucco and wooden walls (fig. 4). In the earliest pictures of the sanctuary, one of Keith's temple gongs rests on the ground below the branches in the chancel, and in services at the church today, this same bell is used for meditative silences during Sunday services. The Japanese gong, the tree branches, the roughhewn bark limbs upholding the Oregon pine roof, the program of impressionistic landscape murals by Keith that line the north wall like inset windows: these elements are all highly atypical features for Christian ecclesiastical architecture, and illustrate how the spiritual and aesthetic experimentation of the San Francisco Swedenborgians could inculcate a form of nature-religion at the turn of the century. An early visitor from the 1890s remarked that "it was as if everything you looked at were the result of natural growth. . . . It tranquilized and satisfied, as nature in the deep woods satisfied."[77]

FIGURE 4. Interior of the San Francisco Swedenborgian Church, ca. 1895. (Collection of the San Francisco Swedenborgian church; photographer unknown; image reproduced courtesy of the San Francisco Swedenborgian Church)

While the asymmetry and roughhewn tree trunks have their antecedents in Swedenborg and Sampson Reed, they might also betray Worcester's familiarity with the vernacular mountain lodge architecture that had been developed in Yosemite Valley from the 1860s onward. Worcester had first visited Yosemite in 1863, and spent several months working for James Hutchings at his hotel (just as Muir was to do later). In 1864, Hutchings renamed the hotel the "Hutchings House," and began expanding on the original structure. He quickly attracted notoriety for building a large room—available for rent—around a 175-foot-tall cedar incense tree. The so-called "Big Tree Room" quickly became a gag and popular tourist attraction, frequently reproduced for stereoscopes and postcards. When Muir began working for Hutchings, perhaps inspired by this architecture that literally brought living nature indoors, he built his own homespun cabin in Yosemite around a stream.[78] As Leslie Freudenheim notes, "[S]eeing Muir's cabin and other Yosemite structures with exposed, untreated wood inside and out, often with bark left on, may have inspired Worcester when he designed his homes and the Swedenborgian church."[79] The San Francisco Swedenborgian church, the Hutchings House hotel, and Muir's cabin all held in common a kind of domestication of wilderness—and in the case of the Hutchings House, even to the point of becoming kitsch: the popular photographs and stereoscopes of the Big Tree Room show bourgeois scenes unfolding in the cramped space, with well-dressed gentleman in suits reading, writing, or playing the guitar, while women in formal dresses sit and read the newspaper, all performing an indifferent obliviousness to the enormous tree trunk in their midst.[80] There is a kind of aesthetic contiguity between the San Francisco Swedenborgian church surprising the urban visitor with its interior wilderness, intimations of "raw nature" in the big city, and the domestic house and hotel in Yosemite, purportedly the locus of the wild, where the interior of the wilderness lodge ends up revealing the city: a couple self-consciously enacts the social conditions of civilization (doffed hat, rocking chair, newspaper, pipe), in spite of the gargantuan tree trunk. This is the same kind of impulse that led others to bore tunnels for roadways and paths through old-growth sequoia trees, a tourist gag that ultimately, in the end, weakened and killed so many of the ancient trees.[81] The aesthetic of the sublime that sequestered wilderness as something cut off, existing outside of culture and human history, could harbor a civilizing violence within it. In his more overtly propagandist writings about the inchoate national park system, Muir tropes on the wilderness of Yosemite, Yellowstone, and elsewhere becoming transformed for the masses into a "pleasure ground," a "summer pleasure park." Yellowstone turns into a "Wonderland, and thousands of tourists and travelers stream into it every summer, and wander about it enchanted. . . . [It is] a grand health, pleasure, and study resort—a gathering place for travelers from all the world."[82] The site of wilderness slides into the cosmopolitan, a global circuit of middle-class, bourgeois relations. But as Solnit wryly notes, many of these spaces that miraculously appeared to Muir and others like "landscaped gardens" were indeed due to centuries of careful forest-gardening and horticulture undertaken by local indigenous populations, who were subsequently forcibly removed, often by violent (if not overtly genocidal) means, from the federal boundaries created by the National Park system.[83]

Like the presence of Kurtz in Joseph Conrad's *Heart of Darkness*, the heart of American wilderness in these cultural productions reveals not the Other of civilization or its antipode, but a kind of extension of (white) suburbia and its prejudices (Solnit, again: the contemporary experience of Yosemite Valley is so often as "a suburb without walls, rather than a wilderness with amenities").[84] While in a remote corner of the Alaskan tundra, stunned by the spectacular colors of the sky and the soil, Muir writes the following, perhaps even with his painter-poet friend William Keith in mind, whose paintings hung like windows onto nature in the Swedenborgian church, and on the walls of Muir's own study back in Martinez: "I was back a mile or two from the shore, reveling in this color-glory, and thinking how fine it would be could I cut a square of tundra sod of conventional picture size, frame it, and hang it among the paintings on my study walls at home, saying to myself, 'Such a Nature painting taken at random from any part of the thousand-mile bog would make the other pictures look dim and coarse.'"[85] It is a striking, signal moment of aesthetic perception where tundra melds into suburban study, further echoing the germination of the Sierra Club in Keith's San Francisco art studio, when the group informally met in a space that was enclosed by Keith's landscape paintings. The intense beauty of the wild takes Muir not more and deeper into some Alaskan "thing" out there in the tundra, in-and-of-itself, but back to the domestic confines of his house, where he cuts off the earth, framing and containing it like a painting.

Muir's wish to hang the tundra sod on his wall and the San Francisco church's own environmental aesthetics distinctly looks toward later earthworks undertaken by artists in the Land Art movement, in particular the tree-based sculptures of Andy Goldsworthy, who has several important installations that lie just a few blocks away from the church in the Presidio park of San Francisco. The sinuous lines of eucalyptus trunks in Goldsworthy's *Wood Line* (2011), or the slender, tapering pirouette of Monterey pines in his *Spire* (2008) directly (though not deliberately) echo the curving Arts and Crafts madrone boughs that hold up the ceiling of the nearby church. Goldsworthy's work has more recently become a major site of consideration for both eco-art historians and religious studies scholars who see his immersive, organic installations as sites of reenchantment.[86] It is a striking propinquity of place, a kind of loosening of discrete moments of time in a way that recalls the flash of Walter Benjamin's *Jetztzeit*, that has the largest public concentration of Goldsworthy's work in North America veritably lying at the doorstep of Worcester's garden church with its profusion of wooden assemblages in its sanctuary.[87]

The San Francisco Swedenborgians, however, did not embrace the kind of earthy pantheism that has been located in Goldsworthy's work, which has more recently become sites for constructive (new) materialist theologies.[88] The church space remains true to Swedenborg's dualistic, Neoplatonic roots, as suggested by the symbolism of the central stained-glass roundel to which the eye is immediately brought, out of the flickering gloom of fireplace and dark wood paneling, located above the sanctuary chancel. Designed by Bruce Porter (William James's son-in-law), an artist who was, like Keith, also a church parishioner, the window shows a flowering bough of a tree. Below the bough a large birdbath with a perched dove occupies most of the frame, and again below the birdbath, three irises poke up. The window is awash in green and purple earth tones that mirror the subdued colors of the wood interior of the church. What becomes immediately apparent when the window is seen in person (and unfortunately not reproducible in photographs) is how the brightest flare of color in the glass is in the pink reflection of the tree blossoms in the water in the birdbath, and not in the blossom itself. The image in the water, in other words, the world it refracts back, seems to be the source of this central color for the tree's blooms, the center of attention in the middle of the roundel. There is no direct iconographic context or antecedent for this window, so the exact meaning intended by Porter or Worcester can only be surmised. Worcester would have been aware of Swedenborg's own iconographic program in the first Latin editions of his work, which Worcester had studied firsthand in the library at the New Church Theological School in Cambridge, Massachusetts. Swedenborg had absorbed the craft of illustration and engraving while staying with different tradesmen's families in Amsterdam and London—the same working-class milieu that later produced William Blake—and Swedenborg subsequently came to create a number of unique ornaments and images for his theological works himself.[89] The image of a fountain in an enclosed garden, with a cherub watering flowers, and a small building at the end of an allée of trees, seems to have been one of Swedenborg's favorite title-page ornaments, as he used it for eight separate theological volumes, including the immanently themed *Divine Love and Wisdom*. Within the alchemical and

Neoplatonic textual traditions with which Swedenborg was familiar, a fountain or a pool of water within an enclosed garden space carries deep iconographic resonance, functioning as a symbol of doctrines of emanationism as well as the "fountain of wisdom" that lay at the heart of so many esoteric teachings.[90] In the Swedenborgian church, the pool of water in the birdbath, with its gash of pink blossom sharper, brighter than the tree itself, evokes this Idealist tradition, and suggests how the aesthetic totality of the church is meant to draw our inner eyes to a spiritual world *within* the natural. In no way can Worcester's garden church be seen as merely a temple to nature itself; it is meant to get beyond the "natural landscape" to the (spiritual) "state of feeling which the landscape produces," as Keith put it about his own aims with his painting. Swedenborgian theology remains unequivocally Neoplatonic and dualist, even as this position is somewhat modified by Swedenborg's immanental tendencies toward a kind of panentheism. "In hell, all worship Nature," Swedenborg darkly warns; "to worship the sun and the moon of this world and bow down to them, signifies a love of self and the falsities that spring from love of self, and it is said that such would be cut off."[91]

Swedenborg's correspondential system depends wholly on the keen distinction between the natural and the spiritual (the latter flowing into, shaping the former), as Wouter Hanegraaff and others have made clear. To conflate the two would be to collapse the foundational structure of his spiritual semiotics. Swedenborg "reintroduces an element of dualism which posits the superiority of spirit over matter," as Hanegraaff phrases it, at a historical juncture when esoteric theories of correspondences were generally turning more toward an animated conceptualization of nature.[92] For an aesthetics based on Swedenborgian correspondences, like Keith's, the purpose of art was to point inward toward this spiritual reality behind the outer "husk" of nature (this was, again, Keith's critique of Muir, that he never got beyond the outer "husk" of the material world). In the series of paintings that Keith made of Joseph Worcester's Piedmont cottage—the home that Worcester had built for himself in 1878 in the unincorporated hills above Oakland—the dramatic rooflines of the house are made to mirror the sloping diagonal lines of Mount Tamalpais across the bay, and the pastured hills that spill down to the bay below: the painted representation of the home "corresponds," in other words, to the topographies of land and horizon, creating a parallelism for the eye between natural forms and the shape of Worcester's bungalow (fig. 5). With this kind of integrated harmony, no wonder, perhaps, that it "came nearer to Jack London's ideal than any [other] house that he ever dwelt in," as London's daughter Charmian later remembered.[93] Contemporary reviews of Keith's paintings of

FIGURE 5. William Keith, *Reverend Joseph Worcester's House, Piedmont,* ca. 1883. Oil on canvas, 12.5 x 24 inches, signed at lower left; inscribed "S.F." at lower right. (Private collection; image reproduced courtesy of Alfred Harrison and the North Point Gallery, Berkeley, California)

Worcester's home seem to have responded to the painter's idealist intentions: "This insight into the reality of nature," enthused an art critic in the *Argonaut,* "by which its ideality is discovered, is the goal which [Keith] has in view. Nature is there as an evidence to the senses; *but through it all may be seen a higher meaning.*"[94]

But, just as our eye in the Swedenborgian church is drawn up to the reflected pink glow in the fountain water of the stained-glass window, getting "beyond" the tree itself, does such spiritualizing idealism end up jettisoning the actual ground and the particularities of a place—and a cognizance of the environmental conditions necessary for a living-with nature that Aldo Leopold would later call a "land ethic"? The Swedenborgian church embodies a great paradox in the preciousness of its nature on display that yet depended on the wholesale destruction of old-growth redwoods and madrone trees. Worcester went to great lengths to procure the madrone trunks for his church, apparently against the wishes of the trees' owners in the Santa Cruz mountains, to whom the trees were "very dear." Freudenheim observes the irony in the church helping to launch an architectural movement that was devoted to bringing nature home, indoors, that yet through its immediate popularity, and subsequent boom of the Craftsman building craze in the Bay Area, led to the rapine destruction of redwood forests along the northern California and Oregon coasts.[95] Worcester further oversaw acquiring the old-growth redwood for the church through his parishioner Mary Curtis Richardson, another painter (and student of William Keith's) who had married a rich Canadian lumber merchant. Worcester was by no means the only one in the Swedenborgian circle to manifest these paradoxes, which were in tension with the conservationist tendencies that Swedenborgian theology could simultaneously enable. Keith first became financially successful as a painter while working under commission for an Oregon navigation and railroad company that wanted dramatic landscapes of the Pacific Northwest for purposes of encouraging commercial development and logging.[96] Keith's ensuing canvases—such as *On the Columbia River* (1869), and *Sunrise, Columbia River* (1869)—feature figures in the foreground evaluating the pink horizon of an undeveloped woodland that unfolds in the morning before them: the images are veritable rosy prospects, and anything but innocent, containing none of the marked ambivalence that characterizes similar commissioned paintings made by George Inness for the Delaware, Lackawanna, and Western Railroad in the 1850s (Inness's *Lackawanna Valley* is discussed in greater detail in chapter 5, as it has become a signal point of reference within American ecocriticism). Moreover, later in the 1870s, Keith began speculating in mining stocks, subsequently becoming very wealthy as a result.[97] The San Francisco Swedenborgian church's beautiful program of the four landscape paintings on the north wall, showing forests of California live oaks and the golden hills and fields of the Bay Area, were painted from a position of privilege and wealth enabled by the most rapacious form of extractive capitalism on the planet. This ambivalence is not that different from Swedenborg's own, if we recall the intrinsic ways that the Swedish mining industry, the larger open-pit iron and copper mine at Falun, fundamentally shaped Swedenborg's wealth, scientific output, and later theology, as chapter 2 has discussed at greater length.[98] The mining technologies propounded in Swedenborg's great *Principia* helped generate the carbon-based economies of the Industrial Revolution, and remain part of the ideological DNA of the Anthropocene.

Muir's consistent attention to trees as beings with even something like the legal status of rights stands in stark contrast to this sort of instrumentalization.[99] His writing, as the new century dawned, increasingly went on the offensive against the "tree killers" in the forestry and agricultural industries, those "wool and mutton men" who were "spreading death and confusion in the fairest groves and gardens ever planted."[100] Muir's fury is at times incandescent, anticipating the later polemics of Earth First! and Edward Abbey (who both in no small measure are deeply marked by Muir's wilderness gospel). In the preceding quote, taken from "The American Forests," which had first appeared in the *Atlantic Monthly* in 1897 (two years after Worcester had built the Swedenborgian church, using so much old-growth timber), the threatened trees are endowed with a "religion" that vanishes under the ax and saw of the woodsman: "[E]very summer thousands of acres of priceless forests, with their underbrush, soil, springs, climate, scenery, and religion, are vanishing away in clouds of smoke, while, except in the national parks, not one forest guard is employed." Throughout his writing Muir grants trees a kind of ontological being and agency of their own, and seeks to understand what they are communicating. Rather than this constituting a very outdated sort of Romantic pathetic fallacy, Muir's openness to trees-as-persons accords with new materialist ethics, as well as the recent cutting-edge research undertaken by the German forester Peter Wohlleben (among others). Muir's attunement to the communicability of trees is borne out by Wohlleben's contemporary scientific findings on a true kind of arboreal semiotics, the ways various species use networks of fungal biochemistry and pollen to reciprocally communicate with one another (if at a much slower tempo than the human).[101] In the end, Muir's locating of the trees in a kind of Uexküllean *Umwelt* diminishes the place of the human, or at least greatly qualifies it.[102] This is clear at the end of a famous moment in the *Mountains of California* where Muir rides out an intense windstorm on the top of a hundred-foot Douglas spruce, listening to the "varying tones of individual trees—Spruce, and Fir, and Pine, and leafless Oak" as they thrashed and bent about. Muir concludes: "We all travel the milky way together, trees and men; but it never occurred to me until this storm-day, while swinging in the wind, that trees are travelers, in the ordinary sense. They make many journeys, not extensive ones, it is true; but our own little journeys, away and back again, are only little more than tree-wavings—many of them not so much."[103]

This equivalizing kinship of humans with trees is also evocative of Swedenborg's own charming device from his *Worship and Love of God* prose-poem, where the animals bear antlers and horns that resemble branches as hereditary signs of their arboreal descent from the tree-life that first hatched out of the primordial egg of the world, birthed by the father sun from the same inseminating life force that produces Adam the first-born. As Muir puts it, again: "[O]ne fancies a heart like our own must be beating in every crystal and cell, and we feel like stopping to speak to the plants and animals as friendly fellow mountaineers."[104]

Herald of a New Gospel

Muir's broad ethical concern for the other-than-human extended beyond an animating regard for the trees; it is also present in his "Stickeen" narrative that recounts his Alaskan adventuring with the eponymous dog. Just as Joseph Worcester's Arts and Crafts "bungalow" in the East Bay, built according to the same Swedenborgian principles of correspondence as the later church, provided a creative backdrop for the drafting of Jack London's classic *Call of the Wild,* so does Swedenborgian theology drape itself over the making of Muir's famous dog story, more or less contemporaneous to the creation of London's most famous naturalist novel. Muir began to write out a plan for his story on the flyleaf of a John Brown volume of stories and essays: "Stickeen," he scrawls, "Like Swedenborg a Herald of a new gospel." The Dr. Brown books had been given to Muir by Keith; both likely appreciated the Scottish physician's evocation of rural life and the landscape of their respective childhoods. Brown was an enormously popular author whose realistic portrayals of human-animal relationships mark him today as a kind of forgotten Victorian-era James Herriot. "Rab and His Friends" and other dog stories in Brown's *Horae Subsecivae* (1858) prompted Muir to recollect his earlier exploits with Stickeen, jotted quickly down here and there in the book, which then culminated with Muir publishing, in 1897, "An Adventure with a Dog and a Glacier," and later on, the expanded text called *Stickeen* in 1909. Both published versions, however, drop all references to the Swedish mystic that first appear in the notes around *Horae Subsecivae.*

The narrative of *Stickeen* is fairly straightforward: in 1880, Muir sets out to explore the icy regions of lower Alaska, and befriends en route a strange, quirky little dog named Stickeen, so-called because of the way the local Stickeen (Tlingit) indigenous peoples had claimed the creature as a kind of "good-luck totem," a "universal favorite; petted, protected, and admired wherever he went, and regarded as a mysterious fountain of wisdom."[105] At the outset, thus, Stickeen is linked metonymically, totemlike, to Muir's encounters with the Tlingit peoples. One morning while camping in backcountry wilderness along the coast, a fierce storm awakens Muir's curiosity, and he sets out into the glacial moraine and steep mountains to better observe the weather (or rather, to absorb it: Muir's witnessing of the storm becomes, typically, an ecstatic, sublime kind of contact). The plucky black dog follows Muir, despite Muir's remonstrance for him to return to the camp, and the two find themselves in dire straits when Muir loses his way among the dangerous crevasses on a glacier. When night begins to fall, the two must leap and slide together over precipitous drops in a desperate attempt to find their way back to warmth and safety. The intensity of their death-defying experience bonds the two mortals deeply, and the text closes with their joyous return to camp, and Stickeen's ultimate disappearance. But the new gospel heralded by the mysterious Stickeen, this good news that Muir claimed—surprisingly, given the brevity of the narrative—was the "hardest thing" that he ever had written, nevertheless remains something of an enigma.

Within its generic parameters of the early twentieth-century wilderness-adventure tale, Muir smuggles in a deanthropocentrizing "ray of relation" (to put it in Emersonian terms) between human and dog. Muir's new understanding of Stickeen after their near-death experience emphatically takes on the aura of religious epiphany: "So hidden before, [Stickeen] was now transparent, and one could see the workings of his heart and mind like the movements of a clock out of its case. His voice and gestures, hopes and fears, were so perfectly human that none could mistake them; while he seemed to understand every word of mine."[106] Muir's sense of Stickeen as "perfectly human" is rendered overtly theological through the poetic epigraph that opens the later novella, a poem by Josiah Gilbert Holland (a figure as once famous, now as equally forgotten, as Dr. John Brown). Muir excerpts the portions of Holland's poem that ponder

> I look into your great brown eyes,
> Where love and loyal homage shine,
> And wonder where the difference lies
> Between your soul and mine.[107]

At the end of the narrative, this "transparency" of Stickeen has become a figurative portal for Muir's expanded, ethical sense of connection to other-than-human beings: "[O]ur storm-battle for life brought [Stickeen] to light, and through him as through a window I have ever since been looking with deeper sympathy into all my fellow mortals"—the mortals being both human and the other-than-human, brought together by a shared condition as earthbound, death-tilting creatures.[108] This chthonic dimension of the story that causes, on the one hand, Muir to decenter the place of the human while, on the other, to expand agency and being to include the other-than-human, strikes the tone of the same theological heterodoxy (if not heresy) that Muir entertained in his private journal while traveling through the Florida Keys in 1867, written during his incredible walking journey from Indiana to the Gulf Coast, when he was twenty-nine years old.[109] Biographers of Muir and environmental historians concur that this meditation, instigated by an illness that nearly killed Muir and his wondering observations on the tropical profusion of life flourishing around him, is a central event in his life, formative in developing his later environmental conscience.[110]

Published only posthumously in 1916 as part of *A Thousand-Mile Walk to the Gulf*, Muir's cogitation on the diminished place of the human in the cosmos probed the limits of theological (and Transcendentalist) orthodoxy. "Cedar Keys" deploys some of the same language that appears much later in *Stickeen*—most strikingly, for example, the phrase "fellow mortals":

> From the dust of the earth, from the common elementary fund, the Creator has made Homo Sapiens. From the same material he has made every other creature, however noxious and insignificant to us. *They are earth-born companions and our fellow mortals.* The fearfully good, the orthodox, of this laborious patchwork of modern civilization cry "Heresy" on every one whose sympathies reach a single hair's breadth beyond the boundary epidermis of our own species. Not content with taking all of earth, they also claim the celestial country as the only ones who possess the kind of souls for which that imponderable empire was planned. . . . Plants are credited with but dim and uncertain sensation, and minerals with positively none at all. But why may not even a mineral arrangement of matter be endowed with sensation of a kind that we in our blind exclusive perfection can have no manner of communication with?[111]

Muir's sensating minerals here are "vibrant matter," his framing of these questions exhibit what Jane Bennett has called "geoaffect": a sensitivity to the fundamental vitality of matter that "is born of a methodological commitment to avoid anthropocentrism or biocentrism."[112]

Muir's wondering about the agential feelings of a rock are also simultaneous, perhaps not coincidentally, to the inchoate field of psychometrics, which flourished between the Civil War and the close of the century. As Dana Luciano has shown, this somewhat odd meeting of geological earth sciences with the telepathic medium of spiritualism (a field diffused with Swedenborg's presence) produced a body of esoteric writing around what she calls, following Michael Taussig, "unthought zones of materiality."[113] Psychometrists like William and Elizabeth Denton attempted to listen, empirically, to the spiritual resonances and acoustical auras of material objects—trees and rocks, fossils, sand and soil—and to telegraph their communicability through the (usually female) body of a medium under a trance-state. As a kind of "enchanted geography," as Luciano puts it, in spite of its bizarre methods, psychometry nevertheless enabled some nineteenth-century persons to record their *feelings* of deep time, and to experience a kind of materialized connection between their bodies and others. "In sensory psychometry," as Luciano puts it, "a certain pushback takes place in the medium's body, so that even as psychometric geology instrumentalizes its specimens, the process of doing so also briefly *things* the human body, making it conscious of sensations that cannot be easily reaggregated into what it means to be a person, or even alive."[114]

When it came to geology, however, Muir clearly preferred his friend Joseph Le Conte's modern modeling to these more popularized distillations of spiritualist earth science. Muir was nevertheless still deeply, if covertly, interested in the paranormal (even if he notoriously interrupted the chicanery of a séance at a friend's house in San Francisco).[115] Stickeen's proximity to Swedenborg might be properly understood when paralleled with Muir's marginalia in the Swedenborg volumes, where it seems Muir's interest in the seer lay more in Swedenborg's purported psychic powers than in the theology itself.[116] The majority of Muir's underlinings in Wilkinson's biography of Swedenborg, for example, signal attention to Swedenborg's purported telepathic powers that later attracted the interest of Immanuel Kant and others. Perhaps Muir was trying to find himself here, and make sense of his own inexplicable experiences of clairvoyance and premonition (including the death of his mother). Whatever the cause or coincidence, Muir's paranormal encounters were essentially moments of communicability—the transmission of some intelligible feeling or thought inexplicably across barriers of time, matter, or space. At its heart, this seems to be the electric transmission at the center of *Stickeen,* where a profound kind of communication—if not communing—takes Muir out of himself into a transformative contact with a dog, that then broadens the radius of his ethical attachments: "I have ever since been looking with deeper sympathy into all my fellow mortals."

Swedenborg's panentheism lay on the fringes of orthodox Christianity (and porously seeped into various spiritualist phenomena, like psychometry), but Muir would have also found his own spiritual inclinations toward the other-than-human corroborated by the indigenous traditions that he encountered. On the same Alaskan trip as when he met Stickeen, Muir also recorded rich conversations and observations of the native peoples who traveled alongside them, including Stickeen tribesmen. Among Muir's traveling companions was Samuel Hall Young, a Presbyterian minister and missionary who was on good footing with many local tribes in Alaska. Muir seems to relish reporting an evening when a Stickeen friend, Kadachan, paused in conversation at the night campfire as a wolf howl broke across the way—"Kadachan puzzled the minister with a question, 'Have wolves souls?' The Indians believe that they have, giving as foundation for their belief that they are wise creatures who know how to catch seals and salmon."[117] Though Muir, then, left no answer (at least in writing) to this question himself, soul-filled Stickeen, with his eponymous name linking him to Kadachan, would seem to echo a reply in Muir's published story, many years later (*Travels in Alaska* was published posthumously, after *Stickeen* had first appeared). This animistic spreading of a soul into the other-than-human was not a view that would have been shared by Joseph Worcester and the San Francisco Swedenborgians; Swedenborg remained too much of an eighteenth-century Rationalist to lose his grip on an anthropocentric hierarchy, and was fairly explicit throughout his theology: believing that animals had souls was simply a "crazy notion," a "hallucination" caused by the deluded love of our own human intelligence; humans are born as animals, and inherently inclined toward evil, "but become human" through spiritual regeneration and recognition that "nature is dead," except when it is seen as an emanation of spirit.[118]

In the end, Stickeen seems to elude the conclusion of Muir's narrative, perhaps in a way analogous to the slippery place of Native Americans in Muir's nature writing as a whole—which certainly remains one of the more contested aspects of Muir's legacy.[119] There is not adequate space remaining to rehearse these debates in environmental history and critical biography here; suffice it to observe that Muir's perspective on indigenous people certainly evolved from one of stereotypical ethnocentric disdain, even racism, to a growing appreciation for native cultures that were under threat from the same civilizing forces that drove Muir himself out into the wilderness. An oft-cited moment in Muir's prose of his purported racism are his earlier observations of the Mono Paiute who lived in Yosemite, and yet "somehow seemed to have no right place in the landscape"; they were "mostly ugly, and some of them altogether hideous. The dirt on their faces was fairly stratified, and seemed so ancient and so undisturbed it might almost possess a geological significance."[120]

Muir's later recollection and writing about Stickeen was an intense experience, coming after his transformative encounters with tribes of the Pacific Northwest, and he uses the present tense for a sense of immediacy at the very end of his dog story, how "it all comes rushing and roaring to mind as if I were in the heart of it . . . I see little Stickeen, I hear his cries for help and his shouts of joy."[121] But then the narrative dispenses with the beloved dog, more or less by the force of a single sentence: "After my work for the season was done I departed for California, and I never saw the dear little fellow again." Muir continues to explain that he anxiously sent letters to Alaska, asking after the dog's fate, and subsequently learned "that in the summer of 1883 he was stolen by a tourist at Fort Wrangel and taken away on a steamer. His fate is wrapped in mystery."

Such opacity, after all the language about the Stickeen as transparent portal, is striking, and it might be a moment where Muir's narrative becomes inadvertently entangled with the larger cultural dispossession of indigenous peoples' lands that Muir's project of wilderness enchantment tended to reify (or at the very least, cloak and obscure). Tourism and colonialism "stole" Stickeen, vanishing the dog from Muir's purview in spite of his intense affection, which are the very same cultural forces that rendered indigenous Americans such as the Mono Paiute as a past obsolescence, severed from any particular futurity of a place like Yosemite Valley (the dirt on the Paiute, who "have no right place in the landscape," "so ancient and undisturbed it might possess geological significance"). This bifurcated vision of wilderness continues to ghost the ongoing indigenous struggles for recognition of sacred sites now deemed federal wilderness conservation areas, even if such land—today—is no longer overlaid with the same spiritualizing "language of things" that Swedenborg's theology had once effected for Muir and the San Francisco Swedenborgians.[122]

5

Homes for Herons

The Eco-Aesthetics of Sarah Orne Jewett and George Inness

> Any spot of country where there was still sky and plenty of trees had a way of waiting quietly around Jewett until she suddenly realized that she had become a piece of it, instead of a foreign substance thrown onto it by chance.
>
> —F. O. Matthiessen

> God all space fills.
>
> —George Inness

The last chapter diagnosed the ways Swedenborgian correspondence and influx abetted the construction of a wilderness mystique in the West. This final chapter will address a different, coterminous reception of Swedenborg back in the East, where the encounters with Swedenborgian theology in the work of the painter George Inness and the regionalist writer Sarah Orne Jewett produced a kind of environmental aesthetic that was localized and numinous, eschewing the more bombastic forms of the sublime that had underwritten John Muir and William Keith's turning of "mountains into bibles" in the California Sierra Nevada. Inness and Jewett did not associate with one another, though they circulated through some of the same northeastern Transcendentalist ambits, and additionally had their work reviewed or published by the same journals.[1] Nor is the characteristic style of Inness's Tonalist landscapes easily correlated with the folksy regionalism of Jewett's fiction; if anything, Inness's dematerialization of locality would seem to jar against Jewett's embeddedness within the communities of rural Maine that characterizes her style, its peppering of argot with pungent descriptions of specific New England folkways (a closer painterly analogue would ostensibly be the crisp coastal landscapes of John Kensett, and others sometimes categorized as "Luminist"). And yet, I argue, Jewett's famous short story, "A White Heron" (long a staple in the green canon of American environmental writing), participated in the same incipient milieu of ornithological conservation that forms a critical backdrop to George Inness's late paintings of water birds. In the 1890s, Inness came to create a sequence of stunning images of herons embedded in diaphanous landscapes—the birds taking on a figurative role normally held in paintings by humans, who so often blur and blend into their surroundings in a kind of atmospheric haze. Though Inness's three different paintings of herons, on their own, do nothing like the direct culture-work and political engagement that William Keith's paintings of Hetch Hetchy inarguably did during the Sierra Club's advocacy for its preservation in the early twentieth century, they nevertheless appear in a context where the conservation and preservation of birds—particularly endangered water birds, like herons—had moved to the fore of an inchoate American environmental imagination. Jewett's short story collection that featured "A White Heron" was published the same year the first Audubon Society was established by George Bird Grinnell, the multitalented anthropologist and natural historian.[2] By the mid-1890s, the bird conservation movement had attracted tens of thousands of members, including Inness's good friend Henry Ward Beecher, as well as Jewett herself, who was a charter member of the Massachusetts Audubon Society that formed in 1896.[3] Inness's three paintings of herons functioned within a larger "ecology of influence," in Daniel Philippon's phrasing, whereby environmental ideas circulated through different social networks at various scales, forming a discursive kind of circuitry that flowed between and beyond works of literature and art, or the activities of a specific group like the Audubon Society.[4]

Both Inness's and Jewett's representations of a spiritualized nature owed a heavy debt to their respective readings in Swedenborg, whose immanental theology provided them with an intellectual and metaphysical framework for their art. This chapter begins by establishing their respective translations of a Swedenborgian "language of things" into their various aesthetic projects, pairing Jewett's debut novel, *Deephaven* (1877), with earlier paintings from Inness's career that either evince his struggle to visualize Swedenborgian concepts or show how his landscapes became entangled with nascent nineteenth-century environmental concerns. Inness's painting of the New Jerusalem—of which only fragments survive—idealizes a kind of being at home and in harmony with nature that echoes, if in discordant ways, the incipient bioregionalism of the seaside community depicted by Jewett in *Deephaven*. After establishing these points of intersection between a matrix of readings in Swedenborg and nascent environmental aesthetics, I will move onto the central concern with the homes for herons limned by Inness's eponymous paintings and Jewett's tale. At the center of both representational strategies are moments of rarified, spiritual perception that blur categories of distinction not only between viewer and the objects perceived, or between figuration and atmospheric abstraction, but between the human and the other-than-human. Whereas this effect creates rhetorical and visual ambiguity, it also sharpens, by contrast, a clear message from Jewett and Inness about what their enframed nature is not: dead mater simply available for commercial extraction and exploitation.

Instead, a higher use—a use beyond the utilitarian calculus of markets and profits—lies in our ecstatic connections with something larger and nonhuman in nature, drawing us out of ourselves. The spiritual valence proposed by these environmental aesthetics chafes against both the laissez-faire economics of the Gilded Age, which saw the natural world primarily in terms of resource extraction, as well as the cornucopic optimism of nineteenth-century Progressivism that blithely assumed advances in technology and science would ensure an endless supply of material goods for all of humanity's needs, thereby ameliorating problems of growing social inequity. Such ideologies persist under the guise of our present neoliberal orthodoxies about global capitalism; to recover Inness's and Jewett's spiritualized representations of regionalist bird life is to activate an alternative future in the past, a road not taken that can productively unsettle our present. That they respectively chose to portray an endangered species is a further opportunity to graph the emergence of animal extinction as a distinct representational strategy shaped by nineteenth-century cultural anxieties. As with the battle over Hetch Hetchy addressed in the last chapter—where Keith's landscapes and Muir's heated rhetoric attempted to intercede against the federal law—the herons of Inness and Jewett, imbued with their Swedenborgian haloes of the numinous, indicate how a Swedenborgian theology could lend itself to the articulation of a nature imperiled, a place with beings and biota not resplendently located in an idealized spiritual world but materialized, vulnerable, and fragile.

From Tree Stumps to Green Cities

Of all the nineteenth-century American writers and artists who gravitated toward Swedenborg's teachings, Inness's translation of Swedenborgian theology into an aesthetic theory and innovative landscape praxis stands out as a singular achievement. Though his canvases do not simply distill doctrines in a denotative or illustrative way, his obsessive focus on using the landscape genre to distort and complexify perceptual processes is fundamentally entangled with his readings in Swedenborgian correspondence theory and its cognitive implications, as Rachel DeLue and Sally Promey have formatively established. Inness is somewhat unique among American landscape painters, DeLue observes, in articulating his wish for having been a philosopher, rather than a painter; his paintings accordingly function as metaphysical experiments that test out various hypotheses on the spiritual dimensions of color, space, and form.[5] Inness drafted "piles and piles of manuscripts" about theology, and although these were all destroyed by a house fire in the twentieth century, enough of Inness's extant publications and interviews with art critics show him not merely as a "card-carrying" Swedenborgian artist like William Keith, but as a thinker who transformed and exceeded Swedenborg's original eighteenth-century Neoplatonism through depictions of nature that are characterized by a hazy blurring of distinctions between the natural and the spiritual.[6] Inness's late canvases show less the nowhere of a Swedenborgian spiritual world, and more of, as Emerson earlier put it, the spirit "as matter reduced to an extreme thinness: O so thin!"[7] Art historians have consequently characterized Inness's sense of nature as pantheist, to a certain extent—his paintings show a nature that was "'an expression . . . of the Godhead' immanent (or homogenous) with divinity."[8]

One of Inness's pre-Swedenborgian paintings, created before he was baptized into a Swedenborgian congregation in 1868, has become a focal point for ecocriticism and cultural historians concerned with the history of environmental consciousness. *The Lackawanna Valley* (ca. 1855) has become iconic within cultural criticism, as its well-known depiction of a deforested landscape, desultorily gazed upon by a reclining figure among tree stumps (fig. 6), has graced the covers of seminal contributions to the field, including Leo Marx's *The Machine in the Garden* (1964), a foundational work for American studies, and Angus Fletcher's *New Theory for American Poetry* (2004), an important contribution to ecopoetics and ecocritical theory.[9] Originally commissioned by the Delaware, Lackawanna, and Western Railroad Company, the painting would seem to distill the mid-century taste for depicting what Marx (and art historians following his lead) have termed "the middle landscape": a scene of

FIGURE 6. George Inness, *The Lackawanna Valley*, 1855. Oil on canvas, 33 7/8 x 50 3/16 inches (86 x 127.5 cm). (Gift of Mrs. Huttleston Rogers, National Gallery of Art, Washington, DC; photograph courtesy National Gallery of Art under Open Access License)

transition, of transforming unsettled wilderness into pastoralized civilization; in other words, that quintessential making of America instilled in the ideology of Manifest Destiny.[10] This panoramic adaptation of the American frontier became a benchmark for the painters associated with the Hudson River School, exemplified by Thomas Cole's famous *View from Mount Holyoke, Northampton, Massachusetts, after a Thunderstorm* (popularly known as *The Oxbow*, ca. 1835–36), where the sweep of the Connecticut River unfolds a dramatic transition from storm-swept wilderness to deforested farmland. For Cole and other Hudson River painters, an explicit, moralizing dimension was encoded into the landscape's parameters that privileged the space of untrammeled wilderness, and lamented its loss. The early reception history of these paintings in the 1830s and 1840s, as David Schuyler has shown, is interwoven with burgeoning middle-class leisure tourism and varied attempts to protect rural spaces from (over)development.[11] By the time Inness was commissioned to create a landscape that would celebrate the accomplishments of the Lackawanna Railroad Company in Pennsylvania, however, the middle landscape had transformed its pastoral overtures into a mode that easily accommodated the antinomies between technological progress and a vision of the American land as edenic wilderness—a perceptual apparatus that was both nostalgic and ambivalent, as Marx observes, "a look-both-ways kind of native progressivism."[12] Inness's famous painting has accordingly been read as an attempt to harmonize the effects of industry and civilization on nature: the train engine's smoke parallels the chimney smoke that is rising behind the church steeple on the horizon, for example, suggesting a contiguity of relation between the world of secular commerce and that of the sacred.[13] The viewer's attention is drawn to the middle of the canvas by a triangulation of forms between church, train, and engine roundhouse. The casual, reclining figure in the foreground recalls the leisured stance of Arcadian shepherds in the pastoral mode—only here, instead of sheep, we gaze with this *Rückenfigur* upon tree stumps and train tracks, a very modern form of "loafing" and "leaning" perhaps evocative of Walt Whitman's desultory voice in the "Song of Myself" from *Leaves of Grass* (which appeared the very same year Inness probably painted *The Lackawanna Valley*).[14]

Though *The Lackawanna Valley* has been subsequently read as a significant work in Inness's oeuvre and the larger evolution of American landscape painting, Inness seems to have placed little value on this painting himself. It disappeared in subsequent decades, and Inness, bizarrely, randomly happened to stumble onto it in a curio junk shop in Mexico—he purchased it back, by his own admission, simply "for old time's sake."[15] Inness's own preference was for his later, more daring experiments with paint that had passed through his Swedenborgian alembic. Nevertheless, *The Lackawanna Valley* can be seen as articulating a problem that the later Tonalist canvases would continue to explore: a wish for a living in harmony with nature, and a kind of unification between selves and the natural elements around them. *The Lackawanna Valley* seems less a triumphalist narrative of Manifest Destiny (though it certainly registers this mid-century nationalist fantasy) and more a kind of representation that organicizes the human and the technological, undoing the nature-versus-culture dialectic so intrinsic to the wilderness aesthetics of the Hudson River School that Inness (at times) worked against. As Marx was one of the first to point out, Inness's painting is akin to Henry David Thoreau's ecstatic celebration of the railroad bank that closes his *Walden* (first published the year prior, in 1854)—both "seem to say there is nothing inorganic."[16]

Inness was certainly drawn to the anthropological dimensions of Swedenborg's metaphysics, his Neoplatonic view that saw something human in every aspect of the cosmos: "all the things that have been created reflect the human in some respect."[17] Chapter 2 of this book argued that when pushed to its limits, Swedenborg's anthropocentrism exceeds the very boundaries it establishes through its refraction of humanity into all things—this is the prospective posthumanist element that imaginatively populates Swedenborg's reflections on extraterrestrial life on other planets (what is the human, when it takes the form of crystalline globes, walks on all fours, or voices "eructations" from the stomach while ambling around on the surface of the moon?).[18] Inness eschewed the Hudson River School's orientation toward wilderness, avowing a preference instead for what he called "the civilized landscape," where the human is resolutely present. In an interview from 1878—well into his Swedenborgian phase—he declared:

> Some persons suppose that the landscape has no power of communicating human sentiment. But this is a great mistake. The civilized landscape peculiarly can; and therefore I love it more and think it more worthy of reproduction than that which is savage and untamed. It is more significant. Every act of man, every thing of labor, effort, suffering, want, anxiety, necessity, love, marks itself where it has been. In Italy I remember frequently noticing the peculiar ideas that came to me from seeing odd-looking trees that had been used, or tortured, or twisted—all telling something about humanity. American landscape, perhaps, is not so significant; but still every thing in nature has something to say to us.[19]

Inness's sense that "every thing in nature" speaks continues the Transcendentalist project inaugurated by Ralph Waldo Emerson in *Nature* (1836), and earlier by Sampson Reed—to locate not just any "language of things," but a kind of sacred semiotics in the experience of nature itself. This pushes, as noted, the landscapes painted by Inness, William Keith, and others to function simultaneously as images that are to be both seen *and* read (or heard), like a book or a voice.[20]

This kind of duality is apparent in what is Inness's most overtly Swedenborgian painting, *The New Jerusalem* (1867), part of his larger allegorical series he entitled *The Triumph of the Cross*. The paintings of this tripart series—of which only one, *The Valley of the Shadow of Death*, remains wholly intact—were created specifically under the auspices of a very generous grant given to Inness (in 1867) so that he might illustrate his doctrinal ideas and visualize variations specifically on a Swedenborgian theme.[21] Using John Bunyan's classic *Pilgrim's Progress* (1678) as a kind of paratext, the narrativity and loose allegorical structure of the three resultant paintings are atypical of Inness; and yet, they were perhaps his most popular works of art within his lifetime, with *The Valley of the Shadow of Death* viewed by more people than any other Inness painting to date.[22] The series is also remarkable for the accompanying textual apparatus that went with the original 1867 exhibition. Penned by Inness, the prescriptive brochure informed visitors how they were to see (and read) the landscapes, evincing both Inness's wish for his landscapes to inculcate forms of spiritual perception, à la Swedenborg, but also his preoccupation with the cognitive dimensions of visuality itself. It was long believed that the final work in this series, *The New Jerusalem*, had been destroyed by the catastrophic collapse of a ceiling and tower at Madison Square Garden in 1880, where the painting had been on display.[23] Thanks to deft curatorial sleuthing undertaken by Inness scholar Michael Quick, we now know that Inness appears to have salvaged the damaged canvas and turned it into at least three separate paintings. These three fragments were finally reunited in 2004 at the Walters Art Museum in Baltimore, where the "lost" *New Jerusalem* was partially restored and put back together, like the pieces from a damaged jigsaw puzzle.[24] In the three remnants of the painting, we glimpse Inness visualizing Swedenborgian eschatology, and Swedenborg's postapocalyptic reading of the New Jerusalem as the imminent unfolding of a new spiritual age. Within a broader spectrum of nineteenth-century art tinged with Swedenborgian ideas, Inness's *New Jerusalem* stands on par with William Blake's *Jerusalem: The Emanation of the Giant Albion* (1804–20) as an idiosyncratic use of Swedenborg's apocalyptics for an innovative instantiation of the Christian New Jerusalem into a historical present: as London becomes the ground to make sense of Blake's *Jerusalem* ("in England's green and pleasant land"), so is the middle landscape of America's transforming frontier the proper context for locating Inness's eponymous painting.[25]

In what is now called *The Valley of the Olives* (fig. 7), the largest surviving fragment of the original, the honeyed and hazy landscape pushes beyond the urban pastoralism already present in *The Lackawanna Valley*. The shimmering architecture of Inness's New Jerusalem rises seamlessly from the green forest and fields around it, so that it is hard to tell where the diaphanous clouds begin and the buildings end (there is the further suggestion of a fantastical cityscape in the clouds on the horizon when this painting is paired with the other remnants of the canvas). In all three of the fragments, small splashes of color suggest human figures ambulating through the landscape, blurring into their surrounds. If *The Lackawanna Valley* displays an ambivalent take on industrialization and deforestation, *The New Jerusalem*, by contrast, depicts a verdant fusion of nature and culture, with the buildings subsumed by atmospherics of golden clouds and lush trees; contemporary reviewers took note of the "exquisite tones of green."[26] As Sally Promey has further demonstrated, the imagery in *The New Jerusalem* must be seen within Inness's larger project to depict states of Swedenborgian spiritual regeneration, and that even the color scheme would seem to parallel Inness's coterminous statements in New

FIGURE 7. George Inness, *The Valley of the Olives*, 1867. Oil on canvas, 76.3 x 114 cm. (The Walters Art Museum, Baltimore; photograph courtesy of the Walters Art Museum)

Church (Swedenborgian) periodicals about the spiritual valence of particular colors and tones; *The New Jerusalem*, in other words, is not only illustrative of Bunyan's *Pilgrim's Progress*, but reveals the landscape as an unfolding Swedenborgian drama of consciousness and spiritual perception.[27]

Inness stated as much in the accompanying exhibition brochure that instructed how visitors were to view the landscape: "This picture represents a state of peace and rest, after the states of depression and elevation to which the soul has been subjected since leaving the natural light. I have endeavored to convey this thought by a landscape, *where no part is left uncultivated, but all is made subservient to the pleasure and happiness of its residents*."[28] This "subservience" of nature to its human inhabitants echoes the Adamic fiat of the second chapter of Genesis ("be fruitful and multiply, and fill the earth, and subdue it"), underscoring how Swedenborg's schema of regeneration, whereby the "natural" part of the soul had to be sloughed off in order to become spiritual, amplified landscape's anthropocentric predilections. Nevertheless, the contrast between the tree stumps and smoke of the earlier *Lackawanna Valley* with the golden, arboreal hues of the later *New Jerusalem* couldn't be stronger: Inness has envisioned here a futuristic green city, perhaps an allegory partially shaped by Swedenborg's frequent figuration of spiritual regeneration as an organic process akin to the dispersal of seeds, or morphologies in the animal and plant kingdoms. The painting, like Swedenborg's theology, imagines the coming of the New Jerusalem as both entailing the individual transformation of consciousness and the according utopian renewal of society as a whole, of humanity living together and in harmony with the natural world. Some art historians, such as Barbara Novak, have read the tree stumps in *The Lackawanna Valley* as disrupting its bucolic affirmation of industrialization, that they rather register an unease, if not elegiac tone; whatever Inness's original intensions for that commercial commission would have been (and his later diffidence about the canvas should not be overlooked), the shimmering, green tones of *The New Jerusalem* show less the urban pastoralism of the earlier painting and a more ecocentric kind of relation between the human and other-than-human, living together in an environment that undercuts the binary between nature and culture.[29]

By the end of the 1860s, perception of the environmental consequences of deforestation was becoming more widespread across the United States. Hudson River School painters like Thomas Cole had been bitterly complaining of the "ravages of the ax" since the 1830s,[30] but the consequences of overlogging throughout the Eastern Seaboard had become so endemic by the middle of the century that writers like James Perkins Marsh were galvanized into action. Prompted by concern over the disappearance of trees in upstate New York and New England, Marsh's *Man and Nature; or, Physical Geography as Modified by Human Action* (1864) became the first truly ecological work of natural history written in the United States, and a key harbinger of later concerns for sustainable agroforestry—Mark Stoll notes it may be "the single most powerful and influential work in the history of international conservation."[31] Marsh forcefully wove together nascent tree science with an attention to public laws (or lack thereof) and how they could protect forests' integral role in regulating healthy soils and what we would call, in today's language, arboreal ecosystems. Deforestation leads to desertification, Marsh (correctly) opined, and if America wasn't careful, its future landscapes would look more like the desiccated, harsh terrain of Greece and the Middle East, where centuries of unregulated woodcutting had permanently altered the climate. "It is evidently a matter of the utmost importance," Marsh wrote with his sense of urgency, "that the public, and especially land owners, be roused to a sense of the dangers to which the indiscriminate clearing of the woods may expose not only future generations, but the very soil itself."[32] In spite of the book's armature of natural science (derived from Alexander von Humboldt, among others) and charts and statistics, Stoll further observes how Marsh's stance is arguably inflected with mid-century Congregationalist theology, particularly in its Calvinistic moralizing that sounded biblical echoes—"man has too long forgotten that the earth was given to him for usufruct alone, not for consumption, still less for profligate waste."[33]

Inness may have been cognizant of Marsh's book, if not the broader concerns it raised. *Man and Nature* was reviewed in Swedenborgian periodicals that Inness is known to have read (and published in), in addition to the other magazines that Inness probably consumed as part of his voracious intellectual digest. This is not to argue for a causal link between Marsh's text and the forests of Inness's *New Jerusalem*, but to posit that the painting—created four years after *Man and Nature*—might share in a Zeitgeist around a newfound concern for trees and forests in the wake of an unbridled development that had reached unprecedented levels in the 1860s. *The New Jerusalem* is such a strange, striking landscape because of its fusion of futuristic city architecture with so much imagery redolent of the bucolic pastoral. This combination was certainly an irritant for some of its nineteenth-century viewers, who complained that the painting was "an awkward endeavor to precipitate an iron-blue New Jerusalem, of Hottentotish architecture, upon an unprepared and unsuspecting New England farming land."[34] Inness had imploded the narrative expectations of the genre of the middle landscape that depended on a dialectical construction between nature and culture.

As a futuristic green city, Inness's *New Jerusalem* more directly anticipates the later urbanism and civic projects of Daniel Burnham, the Chicago architect and designer primarily responsible for the layout of the 1893 World's Columbian Exhibition in Chicago, and that city's later 1909 master plan for urban renewal. A cradle (and creedal) Swedenborgian, Burnham had attended New Church educational schools in New England where he befriended Joseph Worcester, shortly before Worcester's pivotal move to the West Coast, and Worcester's central, later role in the swirl of environmental aesthetics and conservationist politics surrounding the San Francisco Swedenborgian church, as chapter 4 has more fully addressed. Worcester and Burnham remained lifelong friends, and the wide penumbra of Worcester's Arts and Crafts influence fell on Burnham in Chicago and elsewhere. Burnham frequently visited Worcester in the early twentieth century during Burnham's extended stays in San Francisco as he developed a comprehensive plan (never executed) for the West Coast city. The older Swedenborgian's disquisitions on doctrine colored Burnham's incipient ideas for the so-called Chicago Plan.[35] Widely regarded today as the first comprehensive modern city plan in the United States,

Burnham's totalizing vision attempted to integrate monumental architectural symmetry with an interconnected system of green parks, rivers, and recreational areas that, when taken in toto, unified urban centers with the suburban fringe into a single metropolitan entity.[36] As Irving Fisher puts it, Burnham's plan "is a massive blueprint for the ideal city derived from Swedenborg's eschatological writings."[37] Fisher and other scholars, most recently Kristen Schaffer, have further detailed how Swedenborg's tripartite scheme of the heavens (as celestial, spiritual, and natural realms) led Burnham to layer three overlapping zones in the streets and parkways of the Chicago Plan, as well as an organic view of streets as "arteries" akin to the circulatory flows of spirits through the cosmological figure of Swedenborg's *Maximus Homo,* or "Universal Human."[38] These more esoteric dimensions of symbolism and Swedenborgian allegory seem less relevant than the ways that Burnham absorbed his church's doctrines of use—that truths and thoughts, or even aesthetics, needed to be grounded in the doing of good actions toward the neighbor, most broadly construed.[39] The City Beautiful movement that Burnham helped precipitate was premised accordingly on the belief that beautiful urban architecture (which included substantial parks and greenways) could ameliorate social problems (such as poverty and crime): the individual experiences of beauty by the denizens of an urban area would lead to the larger regeneration of the community as a whole. Burnham's micro-macro dimensions of urban planning may again refract Swedenborg's figuration of the New Jerusalem as both an actual renewal of earthly human communities and also as an event that happens within an individual's regenerated consciousness.

In Inness's few words about his *New Jerusalem* painting, it is clear a similar Swedenborgian multivalence—a kind of visual polyphony—is at work: the painting "represents a state of peace and rest, after the states of depression and elevation to which the soul has been subjected." It is very likely Burnham would have been aware of this particular painting's existence, given the renown of *The Triumph of the Cross* series, and the further overlap of Swedenborgian coteries that Burnham and Inness circulated through. In addition to Inness's celebrated participation in the 1893 World's Columbian Exhibition in Chicago (with fifteen paintings on display), they had several important friends and acquaintances in common, including William Keith and Joseph Worcester. If Burnham's City Beautiful project in the plan for Chicago harbors a kind of utopian didacticism—that individual encounters with urban beauty could regenerate the social whole—it mirrors Inness's aims for his paintings, who felt that his canvases should "elevate" even the commonest mind. "Long before

I had ever heard of impressionism," he wrote in a letter, "I had settled to my mind the underlying law of what may properly be called an impression of nature, and I felt satisfied that whatever is painted truly according to any idea of unity will as it is perfectly done possess both the Subjective Sentiment—the poetry of nature—and the objective fact sufficiently to give the commonest mind a feeling of satisfaction and through that satisfaction elevate to an idea higher—that is more interior than its own."[40] This spiritual didacticism, however, leads to a kind of instrumentalization of nature within Inness's landscapes, as DeLue astutely observes. Though the paintings certainly show a kind of spiritualized immanence, and point toward the lush materiality of objects, in the end, "things in nature become as abstract terms in a logical calculation, and landscape representation itself becomes a science of reason, a model of seeing and knowing all the truths and secrets of the world."[41]

Is there room for the other-than-human within such a rigid aesthetic schema? Inness's late paintings of birds melding into their surrounds suggest some ways we might work toward this problem. As in Swedenborg's theology itself, tension is rife between Inness's pronouncements about art and theology that, on the one hand, subordinate nature to a higher Neoplatonic spiritual world, transcendent within, and yet on the other, model a kind of immanence that runs the opposite direction, toward an earthy pantheism. There is a sense of this in some of Inness's poetry that has survived (published in 1882 in *Harper's Magazine*), as the following excerpt demonstrates:

> Sing joyfully!
> Bright nature lives
> In us.
> Thought, sight, and sound,
> Mind—all are one.

The poem concludes with the injunction to "Sing joyfully! / God all space fills."[42] This last gnomic utterance echoes both Swedenborg's own koanlike pronouncements (in *Heaven and Hell* and the *Divine Love and Wisdom*) how "divinity fills all space in the universe non-spatially," as well as Benedict de Spinoza's extension of the infinite eternal substance (of God, of nature) into all space.[43] If Inness's landscape painting was meant to teach even the "commonest mind" how to see the material reality of a spiritual world, filling all space, a similar aim is established by Jewett in her coterminous fiction, where her urban female characters (stand-ins for the reader) are transformed by their encounters with rural belief systems that allow them to see anew the spiritual in nature, something they had not grasped before, something numinous that comes to set Jewett's craft apart from the other Regionalist and "local color" writers with whom she is often bracketed.

Lighthouses and Queer Cattails

Jewett's absorption of Swedenborgian ideas was primarily mediated by her early friendship with Theophilus Parsons, a Swedenborgian professor of law at Harvard University, and a significant, liberal intellectual within the New Church movement in New England. Parsons had been a member of the fabled Harvard class of 1818 that included William Emerson and Sampson Reed. It was Reed, Parsons, and several others, including Thomas Worcester (the father of the San Franciscan minister Joseph Worcester), who had formed an early reading group to pierce "the mystic maze" of Swedenborg's scholastic Latin; this group organized itself into a society that became later incorporated into the General Convention of the New Jerusalem (the "New Church").[44] When a twenty-three-year-old Jewett met Parsons at a Maine seaside resort, he was already seventy-five years old, but still prolifically sharp with his pen. He had just brought out a new collection of essays, *The Infinite and the Finite* (1872), that distilled various Swedenborgian ideas through a scrim of nineteenth-century Idealism. The book became very important to the fledgling author; she read Parson's *Infinite and the Finite* at least three times, and came to own an additional four other books by Parsons, and a number of (now lost) smaller tracts containing excerpts from Swedenborg that Parsons mailed to her.[45] Jewett did not develop a taste for straight-up Swedenborg, however, as her letters to Parsons complain—"it was all so plain and 'came home' to me so thoroughly when you talked to me and when I read your book. But I tried some of those tracts you gave me and they did confuse me very much and I did not like them half so well."[46] Though she later admits that her readings in Swedenborg were getting "easier and easier"—"I could not understand them because I was not ready I had not lived enough"—Jewett never seems to have acquired any separate volumes by the Swedish mystic, independent of Parsons.[47] Most literary critics concur that Parsons's Swedenborgian influence began to wane when Jewett moved more firmly into the orbit of William Dean Howells (who, by sheer coincidence, was Parsons's neighbor on the same street in Cambridge in the 1870s), and that in her later fiction—especially in what is widely regarded as her finest achievement, *The Country of Pointed Firs* (1896)—Jewett consciously eschewed the moral didacticism that Parsons had encouraged her to develop in her literary craft.[48]

Nevertheless, Jewett's early encounters with the Swedenborgian concepts of influx and correspondence affirmed for her how the "spiritual interpenetrates the natural," as Josephine Donovan puts it, the starting point for an "animistic view" where the transcendental incarnates into the physical.[49] Jewett, accordingly, is not only seen as a critical exponent of the heterogeneous movement we now call literary Regionalism, but as powerfully articulating within that genre's parameters a kind of numinous approach to the natural world, and an incipient ecological ethic that includes the other-than-human.[50] Jewett's vignettes of rural country life in Maine have been fully "greened" into the canon of American environmental writing, even as, arguably, their regionalism forms a kind of allonationalism, in Patrick Murphy's phrasing—a structure of belonging that resists these very nation-building identity formations, reshaping the "imagined community" with ecosystem and environment.[51] The presence of Swedenborg's spiritual world in Jewett's fictional *Deephaven* provides a kind of Transcendental dislocation, as we shall see, that moves the place of the story, its human and other-than-human community members, out of the Regionalism of folksy "local color" traditions (which ostensibly constitute the patchwork tapestry of America, *e pluribus unum*) and into queer kinds of temporalities and belonging, where the dead seem to traffic with the living, and the village of Deephaven appears "not in the least American. There was no excitement about anything; there were no manufactories; nobody seemed in the least hurried," as the narrator relates early in the story.[52]

In the rich body of extant letters between Parsons and Jewett, now held at Colby College in Maine, Jewett's interest in Parsons's exposition of various Swedenborgian tenets dovetails with her soliciting advice and input on the stories that would come to compose the *Deephaven* novel. Donovan seems correct in delineating three elements constitutive to the novel that can be located in Jewett's playing the female ephebe to the avuncular Parsons. The first is her interest in Swedenborg's doctrine of use (as expounded by Parsons), which seems to have legitimated, for Jewett, her sense of purpose in writing fiction: that rather than constituting a self-indulgent foray, it was a way to "be more and more useful in the world," as she wistfully wrote to Parsons. Secondly, Swedenborg's immanental influx charged nature in such a way that it made Divine Love and Wisdom available and accessible anywhere, to anyone, so that "choice can overcome chance," as Donovan puts it.[53] Jewett wrote to Parsons, saying, "I can't tell you how much I have always thought of one thing you wrote of in *The Infinite and the Finite* [*sic*]—that in every event there is a chance for us to learn God's lesson and find good, or a chance for us to lose all this. I see it now for myself. Ah! you never will know how much you have taught me, and done for me. I do not know it all myself yet."[54]

This dilemma of choice, the prospect of turning chance into something like moral agency, is suggested early on in *Deephaven,* when the two main characters arrive in the old ancestral home of Aunt Kate, where the young women will spend their summer, unsupervised and unchaperoned in the remote Maine fishing village. In their favorite room in (dead) Aunt Kate's house, stands an enormous fireplace,

> framed in blue and white Dutch tiles which ingeniously and instructively represented the careers of the good and the bad man; the starting-place of each being a very singular cradle in the centre at the top. The last two of the series are very high art: a great coffin stands in the foreground of each, and the virtuous man is being led off by two disagreeable-looking angels, while the wicked one is hastening from an indescribable but unpleasant assemblage of claws and horns and eyes which is rapidly advancing from the distance, open-mouthed, and bringing a chain with it.[55]

These "two disagreeable angels" in blue-and-white tile may reflect Swedenborg's particular stipulation that every soul is born with association of two angelic good spirits (and two evil spirits) who mediate the psychic balance necessary for free will and the moral agency of choice.[56] With the specificity of the Dutch tiles, Jewett might be further toying with an association to the mise-en-scène of Swedenborg's great spiritual crises in Amsterdam and The Hague, as described in his *Dream Diary.* That sensational posthumous material had recently come into larger public view with the publication of Immanuel Tafel's *Documents Concerning the Life and Character of Emanuel Swedenborg* (1877)—something she could have easily heard about from Parsons.

According to Donovan, Parsons also pressured Jewett to maintain a moralizing allegory in her literary aesthetics—which by the end of the writing of *Deephaven,* she seems to have started to chafe against. "I can see how you are discontented with those Deephaven sketches," she pushes back.[57] The letters further show, as Donovan perhaps too quickly glosses over, Jewett's enchantment with a correspondential language of things. Parsons's writings, though not with Sampson Reed's elliptical elegance and penchant for aphorism, could still bring out the immanental flow of Swedenborgian panentheism in a nice turn of phrase: "[T]he earth we live on, the air we breathe, the insect, the bird, the mammal,—all alike and equally exist and live by efflux of being or of life from Him."[58] Jewett expounds at length to Parsons on her felt connection to this Divine "efflux" in a revealing letter that mirrors later moments of spiritual epiphany in *Deephaven*. Jewett's gendered anxiety at seeming too tomboyish to her mentor is also striking:

> And there's nothing that is much harder than being hindered from having one of my wandering, out-door days. I am so happy when I can be out in the woods or open fields and I hope I shall never outgrow my fondness for such days. I try not to let my boyishness (if it is that) make me rude and unladylike. I don't see why it need do so, and my feeling of real friendship for the hills around here, and the things that grow in summer, is very deep, certainly. There are certain trees that seem to have as much character and much the same kind—as people have, and I know them and am glad to see them again when I have been away. They mean a great deal to me that houses never can.[59]

"In the woods," Ralph Waldo Emerson had counseled in his *Nature*, "we return to reason and faith."[60] In *Deephaven*, the two protagonists, Helen Denis and Kate Lancaster (surrogates for Jewett and her friend Kate Birckhead), "read Mr. Emerson's essays together, out of doors."[61]

Deephaven, like its later counterpart, *The Country of Pointed Firs*, is character-based, rather than plot-driven, as it follows Helen and Kate through the rural seascape of a coastal community in serious economic decline (loosely based on Jewett's own childhood in South Berwick), where they have chosen to spend their summer. Each chapter—most were serialized—features vignettes with the country denizens who have lived off the land and the sea. Helen and Kate's desultory reflections were meant to provide two things for the text's readers, as Jewett informed Parsons in a letter (no doubt seeking his approbation)—"the first to help people look at 'commonplace' lives from the inside instead of the outside—to see that there is so deep and true a sentiment and loyalty and tenderness and courtesy and patience—where at first sight there is only roughness and coarseness and something to be ridiculed. And beside this it seems to me that such a life as I told about in Deephaven is so much pleasanter and more real, than what one calls 'society life.'"[62] Jewett's juxtaposition between urban(e) "society life" and the nature-spirituality embodied in the disappearing folkways of Deephaven functions as a kind of literary strategy to negotiate the nature-culture binary that constitutes so much nineteenth-century American nature writing. Like *Walden* by Henry David Thoreau (whose work was very important for Jewett—she owned nine volumes of Thoreau's writings, annotating many of them), *Deephaven* complicates this easy polarity, as the form of narration itself—stylized by the thoughts and interior monologues of Kate—is ostensibly that of highbrow literature: the kind of fiction that was consumed by upper-class readers of the *Atlantic Monthly* or *Harper's* (and not by the rural types that Jewett's aesthetic translates into literary capital). This is quite characteristic of Regionalist writing as a whole, which operates as a "discursive construction" that interrogates the links between local place and national kinds of identity-formations, shaped by literary markets and canons.[63]

The numinous quality of Deephaven—both its moments of nature spirituality and the uncanny tales of the supernatural that are recounted to Kate and Helen by the old sailors and spinsters of the town—further overlays the regional discursive space of the town with the drama of the secular, and the prospect of literary reenchantment. Early in the text, while Kate and Helen are visiting the lighthouse keeper and his wife on Lighthouse Island, the two girls encounter a gaggle of Boston sophisticates who disembark from their pleasure boat to explore the rocky islet. The lighthouse and its keeper's wife, Mrs. Kew, occupy a central symbolic role in the narrative: the lighthouse at Deephaven is locally referred to as "the Light" (capital L), and Mrs. Kew establishes herself as a repository of folk wisdom and valuable village gossip. But the boatload from Boston can see none of this, and Helen overhears one of the visitors (who has mistaken Kate for a local Deephaven yokel) wondering why she (Kate) doesn't work to get off and away—"why, its so awful quiet! Nothing but the water, and the wind, when it blows; and I think either of them is worse than nothing. And only this little bit of a rocky place!"[64] This deprecatory perspective is the one that Jewett assumed her contemporary literary readers would have had, and by the end of the novel, she had hoped that Deephaven as experienced through Kate and Helen's eyes would rather reveal "so deep and true a sentiment and loyalty and tenderness and courtesy and patience," as she put it to Parsons. The novel establishes a metonymy between its ecospirituality and a mission to bring a rural perspective "back" into the city-space of a skeptical modernity in its very last chapter, shortly before Kate and Helen are to return to Boston, when the two go about collecting various treasures so as to remember their special Deephaven days (this is not long after they have been reading their Emerson, out-of-doors). They pick up two great bunches of cat-o'-nine tails, a New England sobriquet for the marsh bulrush (*Typha latifolia*). "They were much in our way on the journey home," Helen narrates, "but we clung affectionately to these last sheaves of our harvest."[65] The first edition of *Deephaven* accordingly featured a prominently embossed illustration on the cover, drawn by Jewett herself, seemingly in reference to these marsh cattails: a golden sheaf of the marsh plants, bound by a ribbon, a visual indication of the spiritual "harvest" that lay within the book's pages. Emerson had written that he wanted his books to "smell of pine and resound with the hum of insects";[66]

Jewett's novel provides a localized, wetlands-specific Regionalist text, quite literally imprinted with the numinous nature-spirituality of the Maine coast.

The Deephaven lighthouse appears in the middle of the novel (and again at its close) in perhaps the text's most acutely Swedenborgian moment, when Kate and Helen have nestled close by the enormous Dutch tile fireplace, with its angels and dramatic scene of judgment, for a serious late-night colloquy about how "the mystery of this world may be the commonplace of the next," and the lively reality of the supernatural among Deephaven's "simple country folk." Such people, wonders Helen,

> ". . . may have a more complete sympathy with Nature, and may hear her voices when wiser ears are deaf. They have much in common, after all, with the plants which grow up out of the ground and the wild creatures which depend upon their instincts wholly."
>
> "I think," said Kate, "that the more one lives out of doors the more personality there seems to be in what we call inanimate things. The strength of the hills and the voice of the waves are no longer only grand poetical sentences, but an expression of something real, and more and more one finds God himself in the world, and believes that we may read the thoughts that He writes for us in the book of Nature."[67]

The chapter concludes with the girls going upstairs to bed while the "lighthouse lamp burned steadily, and it was the only light that had not been blown out in all Deephaven."

What is further significant about this dialogue and moment of perception is its shared nature between two young women who later avow that they wish to live together like the famous eighteenth-century lesbian pair, the Ladies of Langollen, who had retired to a remote corner of Wales to live together as a married couple.[68] These "grand poetical sentences" of nature at Deephaven that Kate Lancaster finds in the waves are not the same "grand mountain manuscripts" of John Muir, which often signals a solitary (male) observer mastering the sublime landscape, but rather a kind of *communitas* between women-loving-women and the simple natural world that envelops them.[69] If Jewett articulates something like a queer ecopoetics here and elsewhere in her fiction, it may have been further inflected by her engagement with the Swedenborgian tradition. Although Swedenborg's own eighteenth-century position on gender and sexual relationships is ineluctably essentialist and heteronormative, his redeeming of erotic love back into spirituality in works like *The Delights of Wisdom Pertaining to Conjugial Love* (1768) had become a central and catalyzing element to his later Romantic reception—perhaps first signaled, most famously, by William Blake's colorful satire of Swedenborg, *The Marriage of Heaven and Hell* (1793).[70] Swedenborg boldly proposed that sexual desire, transformed into conjugal love, continued to provide full pleasure and desire "for conjunction into one" in the other life, so that conjugal partners appeared to Swedenborg as a single angel; in view of the complex (and sometimes confusing) ontologies of Swedenborg's gendered binaries, nineteenth-century artists who responded to Swedenborg's *Conjugial Love* could sometimes produce a corresponding form of "gender trouble" in their work, splitting the iterative conjunction between biological sex and the cultural performance of gender.[71] In *Seraphita* (1833), for example, Honoré de Balzac uses Swedenborg's angelology to construct a hermaphroditic main character—Seraphitus, or Seraphita—who becomes the object of desire between a man and a woman.[72] In the United States, and closer to the literary cultures that Jewett grew out of, Julia Ward Howe drafted her own Balzac-inspired novel about a hermaphrodite who, like the characters in *Seraphita,* avidly reads Swedenborg in the midst of their gender-bending and love triangulations; recent scholarship on Howe claims that Swedenborgian theology is absolutely essential for making sense of the text's transgressions of heteronormativity.[73] Perhaps because of its radical gender play, Howe abandoned finishing her novel and it was published only posthumously (the text was recovered in 1995 by the scholar Gary Williams, who had pieced together different manuscript fragments that were found in the Howe archives at the Houghton Library).

Jewett almost certainly knew nothing of Howe's hermaphroditic manuscript; she was likely, however, to have been cognizant of both Margaret Fuller and Emerson's responses to gender in Swedenborg. Fuller's important feminist manifesto *Woman in the Nineteenth Century* (1845) had found Swedenborg "sufficiently large and noble" in his ideas, and in her private journals, published posthumously, Fuller slid Swedenborg's love between angels and their unification into one into a legitimation of same-sex desire: "[I]t is so true that a woman may be in love with a woman and a man with a man. It is so pleasant to be sure of it because undoubtedly it is the same love as we shall feel when we are angels."[74] Probably in response to his friendship with Fuller and her theoretical speculations on androgyny, Emerson, in his much republished essay on Swedenborg from *Representative Men,* riffs off of Swedenborg's *Conjugial Love* to offer a countervailing view: "[I]n fact, in the spiritual world, we change sexes every moment. You love the worth in me; then I am your husband: but it is not me, but the worth, that fixes the love. . . . Meantime, I adore the greater worth in another, and become his wife."[75]

Although Donovan proposes that a kind of loose "Swedenborgian theory may have provided Jewett with a way of legitimizing her own homosexual passion," it is certain Jewett never would have found such legitimation within Parsons's Swedenborgianism or that of the more orthodox New Church.[76] Several nineteenth-century utopian experiments that had proceeded under the joined banner of Charles Fourier and Swedenborg occasionally attempted to link Swedenborg's doctrines about angelic sexuality with nonheteronormative relationships; but the New Church organizations and its periodicals were very quick to distance themselves from such endeavors, and condemned the Fourierist-Swedenborg alliances as "a disgusting and detestable free love system" that was nothing but "a plan to subvert the present organization of society."[77]

Ecocritics have long located a vital connection between Jewett's queer identity—Jewett, for many years, was in a so-called "Boston marriage" with Annie Fields—and her environmental aesthetics.[78] Such an approach may help us to make sense of the close of *Deephaven,* which, again, circulates back to an evocation of the lighthouse that has punctuated the text as an indication of the region's rural nature spirituality. In a kind of apocalyptic reverie that imagines the desolate future of the village, fully desiccated by the forces of modern industry that have long abandoned it, and "the few old gentlefolks who still linger will be dead then," the narrator wonders if she and Kate will go back to the village "for the sake of old times." The prose of the novel shifts into a subjunctive mood, using the conditional perfect tense to suspend this future prospect: a fraught temporality in fiction, where the novelistic time of *chronos* becomes *kairos,* as Frank Kermode would put it.[79] The two friends/lovers are (re)united in a moment of imagined spiritualized perception that ends with the lighthouse and the ocean waves as a speaking text. Helen (the narrator) writes:

> I should like to walk along the beach at sunset, and watch the color of the marshes and the sea change as the light of the sky goes out. It would make the old days come back vividly. We should see the roofs and chimneys of the village, and the great Chantrey elms look black against the sky. A little later the marsh fog would show faintly white, and we should feel it deliciously cold and wet against our hands and faces; when we looked up there would be a star; the crickets would chirp loudly; perhaps some late sea-birds would fly inland. Turning, we should see the lighthouse lamp shine out over the water, and the great sea would move and speak to us lazily in its idle, high-tide sleep.[80]

This crepuscular scene that is both past and future, drenched with a desire—to be again with her beloved Kate at Deephaven—that is yet unattainable in a present moment of narrative iteration, is a queer kind of temporality and location, in Judith/Jack Halberstam's sense of time and place. The subjunctive mood suspends the prose into a future that is also a past (yet keenly absent in the present). Its impossibility renders an imagined togetherness that "disrupts the normal narratives of time," as Halberstam would phrase it, working against modern regimes of the clock and calendar that are controlled by "institutions of family, heterosexuality, and reproduction."[81] On this point, Jewett was quite clear: when jocularly asked by the poet John Greenleaf Whittier if she had ever been in love, she responded sharply that she had more need of a wife than a husband.[82] As we shall now see, the intense intersubjectivity of this passage where selves merge into one another and with the landscape (that marsh fog, so "deliciously cold and wet against our hands and faces") also marks the apogee of Jewett's "A White Heron."

Homes for Herons

In the 1880s, the American public was slowly becoming cognizant of the widespread slaughter of bird species—largely caused by the millinery production of fashionable ladies' hats—and a burgeoning alarm, if not panic, began to sound regarding the prospective extinction of certain avicultures. After George Bird Grinnell established the first Audubon Society in 1886, a new kind of civic environmental imaginary emerged that was devoted to broadening public awareness of the plight of endangered birds, and the need for sensible conservation legislation at state and federal levels. Without the advocacy of the subsequent Audubon societies that took off in the 1890s, Congress would not have come to develop and pass the Lacey Act (1900), and the later Migratory Bird Treaty Act (1918), which both established, for the first time, federal legal frameworks for the explicit protection of wildlife.

It is against this dramatic backdrop, the development of a broad awareness of aviculture extinction within the excesses of the so-called Gilded Age, that both Jewett and Inness created distinct representations of water birds—of herons, in particular. Jewett's "A White Heron" appeared in 1886, the same year Grinnell launched the Audubon Society and published the inaugural issues of the *Audubon Magazine*. Inness, in turn, after spending substantial time wintering in Florida and the South, created several paintings of herons, two of which are regarded as among his finest works. Of all the species decimated by the millinery craze for putting taxidermied birds and feathers onto hats, it was the large plumed water birds, such as herons and egrets, that were most threatened by this deadly fashion trend. While it is certainly true that neither

Jewett's short story nor Inness's paintings of herons were intended to intercede directly in support of ornithological conservation, at least not in the same way that, say, William Keith's or John Muir's work participated in wilderness advocacy, their collective representations are nevertheless entangled in a broader cultural mesh that had revealed herons (and their wetland homes) as imperiled, facing the imminent prospect of extinction. Jewett's later involvement in the founding of the Massachusetts Audubon Society suggests the political ramifications of her own aesthetic; while Inness's late canvases certainly refract a kind of Swedenborgian dematerialization of place sublated into a spiritual world, his parallel engagements with the utopian ideas of the radical reformer Henry George provide an alternative locus for Inness's conceptualizations of land, property, and—by extension—habitat. And though Inness himself did not leave extant remarks or comments as to what he thought about public agitation for the protection of birds, several individuals close to him (including his grandson), strongly did. To paint a heron in the 1890s, in other words, was to make a representation of not just any bird, but a highly charged species that had been endowed by the mechanisms of American culture with an aura of (contested) endangerment and risk of extinction. Even outside the parameters of Inness's original intentions, the contemporary reception of pictures of herons would have activated a network of these conservationist associations—rippling within a larger "ecology of influence," to reiterate Philippon's felicitous phrasing. The imaging and visual representation of birds was central to the early efforts of the Audubon Society; Grinnell's "Monthly Birds" in the *Audubon Magazine*, which ran from 1887 to 1888, prominently featured reproductions of color plates from John James Audubon's original *Birds of America* (1827–38). American Romanticism could still provide a viable "usable past" for the environmental imagination at the century's close.

In both Jewett's fictional tale of herons and Inness's ethereal paintings of heron habitat, their creators deployed an immanental and numinous sense of nature, one that blurs and distorts perceptual processes (be they visual or verbal) in order to create a kind of intersubjectivity between observer and bird. The numinosity of these representations provides a kind of strange and fragile otherness that strikes a discordant, if not melancholic tone, against the cornucopic optimism of the era, its market-driven myopia that led to the ultimate extinction of the Passenger Pigeon (in 1914) and the Carolina Parakeet (1920), among other species. As Ursula Heise points out, due to the elegiac mode inscribed into the declensionist narrative of species extinction as a whole (which appears on the horizon of biological sciences and the deep time of natural evolution as an imagined prospect only in the latter half of the nineteenth century), "the cultural logic of extinction discourses lie in their function as powerful tools for criticizing or resisting modernization and colonization; so powerful, in fact, that scientific arguments themselves are inflected by this logic."[83] Not surprisingly, it is not only the secular facts and figures provided by ornithology and the nascent science of biodiversity, but also the reappearance of religion, transposed, in the Swedenborg-inflected work of Jewett and Inness that creates a viable form of aesthetic anti-Modernism. Jewett and Inness both eschew a kind of empirical realism, painting (or writing) instead a form of Regionalism whose compass coordinates are the spiritual world of Swedenborg.

Jewett's "A White Heron" became the title of a short-story collection—*A White Heron and Other Stories*—and as with the embossed cattails of *Deephaven*, the cover of this book featured a flying heron, suggestive of how the eponymous story contained a spiritual harvest from the countryside for the text's urban readers. The narrative is fairly straightforward. The protagonist, Sylvia—a name that immediately evokes the Roman woodland deity, Silvanus—is a nine-year-old girl living with her impoverished grandmother in rural New England. Her parents appear to have abandoned her to seek their fortunes in California. Sylvia and her grandmother's poor but harmonious pastoral life is interrupted by a young ornithologist tramping through the countryside who is shooting and bagging bird specimens for his scientific collection. When he learns that little Sylvia has a deep affinity with "wild creatures," he suspects that she may know where to find the elusive white heron, a rare bird he has recently glimpsed in the area. Sylvia feels a budding attraction to the young man with a gun, and wrestles within herself, for she indeed knows where the heron is likely to be found. Late one night, in the strangest part of the story, while the visitor is still staying with them and sound asleep in his bed, Sylvia wanders out into the moonlit forest in her nightclothes to an old-growth pine, the largest tree around for miles. She climbs it as the dawn begins to come up, and with the sun rising "bewilderingly bright," she catches a glimpse of the elusive heron near its mate and nest, in a moment of transcendentalized perception. She scales down the tree, exhausted, covered in pine-sap, imagining what the young man will say when she reveals the location of his object of desire. But then, in a reversal of narrative expectations, once Sylvia is back at the farmhouse, she refuses to speak:

> Here she comes now, paler than ever, and her worn old frock is torn and tattered, and smeared with pine pitch. The grandmother and the sportsman stand in the door together and question her, and the splendid moment has come to speak of the dead hemlock-tree by the green marsh. But Sylvia does not speak

> after all, though the old grandmother fretfully rebukes her, and the young man's kind, appealing eyes are looking straight and deep into her own. He can make them rich with money; he has promised it, and they are poor now. He is so well worth making happy, and he waits to hear the story she can tell.
>
> No, she must keep silence! What is it that suddenly forbids her and makes her dumb? Has she been nine years growing and now, when the great world for the first time puts out a hand to her, must she thrust it aside for a bird's sake? The murmur of the pine's green branches is in her ears, she remembers how the white heron came flying through the golden air and how they watched the sea and the morning together, and Sylvia cannot speak; she cannot tell the heron's secret and give its life away.[84]

The refusal of language here is also the refusal of romantic and erotic attachment, deflecting the energies of the expected "marriage plot" that generates nineteenth-century fiction into a "queer" kinship between Sylvia and her surrogate mother (an atypical family, but typical in Jewett's fiction that is replete with spinsters, aunts, sisters, and grandmothers all living together—but rarely a healthy or normal heterosexual marriage).[85] To not speak to the inquiring ornithologist is also a negation of the scientific; the handsome figure of the gun-toting birdman alludes to John James Audubon and other nineteenth-century scientific birders who often avidly participated in the wholesale destruction of birds in the name of science. There is great irony in the late nineteenth-century groundswell of bird conservation naming itself after Audubon, who infamously complained that his day was wasted if he did not shoot anything fewer than one hundred birds.[86] In fact, one of the greatest obstacles to bird conservation came from scientific lobbies that defended the killing and taxidermy of bird specimens, and cataloguing them into museums, as necessary for the progress of scientific knowledge. When the head of the American Ornithological Union of Jewett's day—a guild of professional and amateur avian biologists—was asked to attend an inaugural Audubon Society meeting in Washington, D.C., he is said to have coolly replied, "I do not protect birds. I kill them."[87] As Carolyn Merchant points out, even Grinnell's initial salvos for bird conservation appeared in the pages of the hunting magazine *Forest and Stream*, where his pleas for sensitivity to bird-slaughter were ironically sandwiched between the advertisements for guns, ammunition, decoys, and hunting dogs, which aimed to appeal to shooters of waterfowl.[88] Even earlier, in 1884, Grinnell was lamenting in *Forest and Stream* how "from Florida to Maine the bird butchers are shooting, netting, snaring and poisoning, seven days in every week and every week in the season."[89]

How, then, to inculcate care and concern for our "fine feathered friends"? Grinnell's answer was to establish an Audubon Society and monthly magazine that deftly recycled Audubon's earlier *Birds of America* into an effective form of popular science, outside the parameters of nineteenth-century ornithology and its symbiotic relationship to game-hunting. Jewett, in turn, with "A White Heron," models a kind of radical immanence in nature that fuses human and other-than-human; this occurs at the moment of Sylvia's ecstatic perception in the pine tree as the dawn comes, and she finally finds the long-sought-after heron's nest. The narrative voice slips here into the perspective of the tree itself, who "must truly have been amazed that morning through all its ponderous frame as it felt this determined spark of human spirit wending its way from branch to higher branch." This leads up to a paragraph where it seems as if the tree itself guides Sylvia's gaze to where she might see the elusive heron, with the prose slipping from an indirect monologue detailing Sylvia's interior thoughts suddenly into a form of direct address:

> Now look down again, Sylvia, where the green marsh is set among the shining birches and dark hemlocks; there where you saw the white heron once you will see him again; look, look! a white spot of him like a single floating feather comes up from the dead hemlock and grows larger, and rises, and comes close at last, and goes by the landmark pine with steady sweep of wing and outstretched slender neck and crested head. And wait! wait! do not move a foot or finger, little girl, do not send an arrow light and consciousness from your two eager eyes, for the heron has perched on a pine bough not far beyond yours, and cries back to his mate on the nest and plumes his feathers for the new day![90]

The tree appears to become the authorial voice, telling the sylvan Sylvia where to look. This blurring of consciousness—of the narrator's voice into the perspective of the trees, but also of Sylvia's connection to the heron, that "arrow of light and consciousness from your two eager eyes," as Jewett puts it—models a similar state of perception as so many of George Inness's late canvases portray, where figures literally dissolve into their environmental surrounds. What makes Inness's late heron paintings so singularly striking is the substitution for the human figure with that of a solitary bird, aglow with a kind of numinousness.

Of Inness's three extant paintings of herons, two had the title "home" affixed to them: one of which is now in the collection of the Art Institute of Chicago, and another, equally critically acclaimed, at Princeton University. The "home" in the title of the paintings may be a posthumous appellation, however; Inness didn't always immediately name his works, and the painting at the

Art Institute was probably first known as *The Sun's Last Reflection*.[91] All three of the heron paintings probably indicate southern wetlands landscapes—in *Home of the Heron, Tarpon Springs, Florida, 1893* (now in the Gulf Coast Museum of Art, Florida), the presence of palm trees and other foliage clearly indicates the specific context of Florida, where Inness began regularly spending his winters in the sleepy fishing village of Tarpon Springs, from 1887 onward. This decade saw a widespread push for the legal conservation of birds in Florida at the state level—there was a growing concern about the wholesale "slaughter in Florida," as an 1887 article in the *Audubon Magazine* lamented, where "no law and no sentiment protects the birds in that State, and they are butchered annually by tens of thousands."[92] The reason for the Florida slaughter was ostensibly plumage for millinery purposes; a subsequent article in the same magazine entitled "Wholesale Destruction of Birds in Florida" specifically addressed the devastation brought to Florida herons and their rookeries.[93] Some citizens, frustrated by the apathy or outright hostility toward conservation legislation expressed by Florida lawmakers, took matters into their own hands. In an anticipation of the later radical interventions and tree-sits coordinated by Earth First! and other environmentalist collectives, Paul Kroegel, a German immigrant who had arrived in Florida in the early 1880s and had become enchanted with the spectacular bird life he witnessed along the coast, took it upon himself to protect a stand of mangrove trees called Pelican Island that was home to thousands of nesting water birds.[94] Kroegel squatted on the island, living with the birds, keeping watch day and night with a loaded double-barrel shotgun to scare off would-be hunters and feather-gatherers. Kroegel's notorious efforts eventually paid off, thanks to the corollary campaigns of Audubon societies, and in 1903, President Theodore Roosevelt signed an executive order declaring Pelican Island as an official bird sanctuary—the first federal wildlife reservation of its kind, which subsequently became the legislative blueprint for the establishment of the National Wildlife Refuge system.

Coastal Florida, thus, was a kind of ground zero for ornithological conservation by the end of the century, and it is against this electrified context that Inness began painting his herons "at home" in their contested wetland habitats. Though Inness left zero trace of opinion regarding the nascent conservation movement around him, one of his closest friends and patrons, the Protestant clergyman Henry Ward Beecher, amply did. Beecher was a member of Grinnell's first Audubon Society, and lent his famous and influential voice to Grinnell's inaugural magazine issue.[95] Given that Beecher was one of Inness's most important advocates, it seems unlikely for Inness to have been unaware of this pressing concern that Beecher saw as a moral, if not spiritual, dilemma. And if there is something to be further said about apples not falling far from their proverbial trees, Inness's grandson, George Inness Hartley, who had spent parts of his childhood in Tarpon Springs, became a public advocate for wildlife conservation in the subsequent decades after his grandfather's death, for birdlife in particular. In Hartley's nonfiction nature writing, such as *The Importance of Bird Life* (1922), and in his serialized juvenile fiction like "The Last Parrakeet" (1923), he expresses both a wish for better conservation and a lament at the loss caused by extinction. His "Last Parrakeet" reads like a Hardy Boys mystery set against the melodrama of species extinction in Florida: two chummy naturalists, Fred and Paul, are sent to Florida's Everglades by a leading museum in New York City to hunt for the last living remnant of the Carolina Parakeet (famously painted and described by Audubon, likely extinct by 1918 or 1920). Dangerous escapades with wild bears and armed ruffians ensue, and the two young naturalists witness and mourn the rapacious destruction of an egret rookery by "a plume hunter."[96] Thus, within the orbit of Inness's immediate family, we can trace the aftereffects of not only a profound concern for the precarity of other-than-human birdlife, but also its aestheticization: a perhaps not insignificant foil to cast retrospectively back on the hazy, diffuse auras of Inness's heron paintings of the 1890s, made at the same time that Kroegel was famously standing guard over the birds with his loaded shotgun on Pelican Island.

Inness himself clearly gravitated toward more radical politics that critiqued the status quo of contemporary industrial capitalism. He became fully interested in Swedenborg while residing at Eagleswood, an abolitionist military academy in New Jersey. Progressive and ardently antislavery, Eagleswood had a steady stream of Transcendentalist luminaries who visited to speak on topics ranging from spiritualism to women's rights. Emerson and Margaret Fuller were both guests, and Henry David Thoreau had helped survey the estate's grounds. After John Brown's failed Harper's Ferry raid, one of the utopian project's founders, Rebecca Spring, arranged for the bodies of two men who had been executed alongside Brown to be brought to and interred on Eagleswood's grounds. Inness was one of several artists who lived onsite at the academy; others included the painter William Page (from whom Inness first absorbed his Swedenborgianism), and the young artist Louis Comfort Tiffany, who was their student.[97] It is important to emphasize that Inness's keen absorption of Swedenborg took place not within the normative confines of the New Church, but in context of this marked antebellum political and social radicalism and its spiritual heterogeneity.

Well after his later full-fledged conversion to the Swedenborgian New Church, Inness became an ardent proponent of Henry George, the reformist who attempted to fundamentally change American property and tax law. George despaired at the growing inequity of wealth in the postbellum period, and with his *Progress and Poverty: An Inquiry into the Cause of Industrial Depressions and of Increase of Want with Increase of Wealth* (1879), he offered a devastating critique of industrial capitalism, diagramming the breakdown of the idea of societal progress. George believed (perhaps naively) that such disparity could be addressed by reformulating property law and ownership around a new "single tax" system. In reworking an older notion of the commons, George believed no one could, in principle, own land or property; that nature of itself belonged to all. "Nature recognized no claim but that of labor," as John Thomas says of George's theory: "And this meant that no man was entitled to the ownership of anything that was not the fruit of his labor."[98] There was an inherent "right of the community to soil." George's single-tax solution was to enact a confiscatory tax on any rents drawn from the ownership of land, property, or natural resources, which ultimately would prevent extortive inflation of rents and speculation on property values, problems that George saw as the sinister causes for the explosion of postbellum urban poverty and boom-and-bust economic cycles (including the California Gold Rush, whose frenzied speculation had originally drawn George to the West Coast). Eventually, George predicted, the single-tax system would lead to the transfer of all privately held land (back) into the public domain, and the making of a new kind of public commons that would forever destroy the pernicious effects of land monopolies. George's reframing of land as a public trust, owned in common, manifests some of the same nineteenth-century intellectual currents that led to the establishment of a national park system, as well as the early flourishing of alternative intentional communities that tried to dispense with private property, such as the various Fourierist associations that had burgeoned before the Civil War (of which Eagleswood in New Jersey was a direct descendant).[99]

There was enough utopian millennialism in *Progress and Poverty* to accord with Swedenborg's postapocalyptic prognostications that a new golden age of spirituality was soon to come, rejuvenating (or replacing) the "old" Christian church with a new dispensation of faith, signified by the vision of the New Jerusalem descending to earth like a bride "adorned for her husband," as Swedenborg had exegeted in the Book of Revelations. Within the fold of the Swedenborgian church circles in New York, a pro-Georgist group quickly formed, and the "New Church Single-Tax League" began publishing a short-lived periodical, *The New Earth,* which made an attempt to link Georgist economics with Swedenborgian theology. Inness was an early enthusiastic supporter: he would interrupt the construction workers who were expanding his family home in Montclair, New Jersey, to eagerly read aloud pages from George's *Progress and Poverty* and extemporize on George's theories that would transform labor and the capitalist system. Along with Swedenborg, according to Inness's son, the single-tax reform was one his father's grand "pet theories."[100] In 1890, Inness delivered a public lecture at a dinner held in Manhattan in honor of George, and the address's subsequent publication constitutes the last written text by Inness that has survived. The speech shows Inness's thorough commitment to related progressive causes, such as imposing an eight-hour workday, while directing no little fiery rhetoric toward the "capitalists and groping monopolists" for creating a veritable "slave labor" wage system in the battle for daily bread.[101] There's nothing explicit about conservation in Inness's speech per se, although Inness does sound an Aldo Leopoldian note when it comes to George's concept of land: "At present there is one principle, a commercial principle, I mean a financial principle, that is universally accepted, that is that a thing that will not pay for itself must be discontinued. Society at this present day is not self-supporting: instead of being self-supporting it is supported by the individual. The individual has certain rights to that which he pays for. *That which he does not create belongs to the community, and that is land.*"[102]

Leo Mazow is one of several art historians who have convincingly argued that George's ideas colored Inness's aesthetic theories and painting praxis, affecting how Inness viewed his proprietary rights over his own paintings that had been bought and sold, and his specific choices to depict scenes of labor in the 1880s.[103] None of those contexts seem relevant, however, when we turn to the late heron paintings, which show neither an overt concern for working Americans nor an explicit problematic with property or land. Responses from art historians have rather amplified these canvases' Swedenborgian concern to "resolve theology into scientific form," as Adrienne Bell puts it.[104] With *The Home of the Heron* (fig. 8) Judith Barter also notes that although Inness may not "use any explicit Swedenborgian symbols . . . the seamless blending of the composition does suggest one of the religion's basic tenets: the unity and harmony of the universe reflect God's presence."[105] The brilliant smoky oranges indeed create a physical numinosity, sharpened by the verticality of the tree trunks, and the diaphanous blurring of foliage: the landscape, quite literally, is aglow.

Given the preciousness of the bird species he chose to make the central, grounding silhouette in the canvas, slightly off-center to the left, Inness's herons would seem to register, or at least indirectly signify, the general cultural

FIGURE 8. George Inness, *The Home of the Heron,* 1893. Oil on canvas, 76.2 x 115.2 cm, signed, lower right: "G. Inness 1893." Edward B. Butler Collection, 1911.31. (Art Institute of Chicago; photograph courtesy of Art Institute of Chicago under Creative Commons Zero Public Designation, CC0)

discourse of bird rarity and extinction afoot in the 1890s. In Jewett's "A White Heron," it is the numinous otherness of nature, imbued with Swedenborgian overtones of correspondence and influx, that provides a stark contrast to the unctuous, gun-toting naturalist: a figure who seems to stand in for modernity and the secular with his Linnaean wish to classify and categorize his bagged dead specimens. It is to this that the protagonist Sylvia refuses language. Inness's beautiful, glowing painting would seem to provide a similar alternative to a dead and disenchanted world, one that was easily instrumentalized by those cunning "capitalists and groping monopolists" that Inness railed against in his speech for Henry George.

In her groundbreaking study on the generation of species extinction as a distinct form of cultural narrative, Heise argues that the representation of endangerment and extinction of animals often relies on older tropes of the elegy and tragedy. As much, if not more, than a matter of statistics provided by the natural sciences, the declensionist arc of species extinction also traces an aestheticized lament about modernity—more specifically, it inscribes a kind of melancholia (and nostalgia) that serves as proxy critique of both colonization and modernization, as has been noted above. The extinct and the endangered operate as synecdoches, addressing deeper cultural dilemmas. This could be said to begin with the first self-reflexive accounts of the demise of the dodo on the island of Mauritius, perhaps the first casualty of modern anthropogenic extinction (the last living species of the bird was seen in 1662). From the eighteenth century up into the present, the dodo has stood as "an icon of regret for those dimensions of nature that are destroyed by modernization and colonization."[106] Heise's sophisticated argument, however, makes no further inferences about modernity's constitutive loss of religion (another kind of declensionist narrative, what Charles Taylor has called "subtraction stories"), nor of the relation between the natural sciences, particularly post-Darwin, and the advent of the secular.[107]

The spiritualizing vectors of Inness's late landscapes, at any rate, key a mode of visual reenchantment, modeling a perceptual blur between figures, be they birds or humans, and their surrounding environs: a kind of haze of colors akin to the literary effects of that central passage from Jewett's "A White Heron" where the narrative voice flows between tree, girl, and bird. Inness wanted to make his viewers struggle to see the reality of a spiritual world that he believed to be numinously flowing within it. As Rachael DeLue has so cogently demonstrated, Inness thought his paintings could inculcate a new mode of spiritual vision, like Swedenborg, and he deliberately made his work correspondingly "difficult," distorting expected pictorial modes to jam and reroute a viewer's perception.[108] Perhaps one task of this regenerated sight was to recognize the precious sanctity of birdlife as an immanence of the Divine in the natural world. Undoubtedly, as beautiful as Inness's late landscapes certainly are, they also exude a distinctly elegiac quality—caused, in part, by the preponderance of crepuscular scenes of moonlight or evening, the diminishment of light. In this twilight of nature, Inness has painted a mood that echoes the melancholia of extinction that Heise has diagrammed as a distinctly modern cultural phenomenon. As Rebecca Bedell's and Wanda Corn's scholarship has insisted, Inness's late landscapes would seem to distinctly mourn a transition from rural pastoral to suburban and urban, imbued with a sentimental nostalgia for a passing world effaced by modernity.[109] "If we were to characterize in a word the feeling created by viewing such works," writes Corn, "it might be 'loss'—a pervasive though understated melancholy which these artists felt as their familiar world succumbed to alien and mechanical forces."[110] Inness was certainly not sanguine about the human race in the 1890s, a decade of intense jingoistic nationalism; his speech for Henry George began with the salvo that "we hear a good deal of talk now about the goodness of the human heart; I deny it en-toto."[111] As Heise elaborates, melancholia's refusal to provide closure on grief (in the classical Freudian distinction between grief and melancholy) can provide a distinct form of political affect through its irresolution and openness; "[M]elancholia is not only a denial of the loss," Heise says, quoting Catrina Mortimer-Sandilands, "but also a potentially politicized way of preserving that object in the midst of a culture that fails to recognize its significance."[112] If *The Home of the Heron* visualizes a cultural melancholy not only of a prospective loss of herons, but also of the evanescence of religious faith, its hazy deployment of a countering Swedenborgian spirituality also opens to a kind of alternative future, not indistinct from the golden atmospherics of Inness's earlier *New Jerusalem* painting—an as-yet unrealized futurity where birds and their homes would stand fully protected, in coexistence with humanity, like that suggested by the green forests and fields of *The New Jerusalem*, which stand in harmony with the visionary city architecture around them. As the twentieth century commenced and legal regimes finally created viable conservation legislation, this millennial wish for God's kingdom coming-on-earth would increasingly acquire the explicitly green dimensions of the ecotopia.

Coda
Johnny Appleseed, Redux

We are born believing. A man bears beliefs, as a tree bears apples.
—Ralph Waldo Emerson

Long after Johnny Appleseed's dispersal of apple seeds and Swedenborgian fascicles on the nineteenth-century frontier, the memory of this early American Saint Francis of Assisi continues to be a potent symbol for the American environmental imagination. From the singer Pete Seeger to the poet Vachel Lindsay, to recent graphic novels by the artists Paul Buhle and Noah Van Sciver, as well as fiction by Richard Powers, Johnny Appleseed remains a locus for articulating our desires and fears about our relationships with nature and the other-than-human.

Appleseed's Swedenborgianism is part of his enduring weirdness, the kind of Americana that thrives in its liminal relationship to mainstream orthodoxies—be they cultural, economic, or religious. For the bright and bold cover of Buhle and Van Sciver's *Johnny Appleseed: Green Spirit of the Frontier* (2017), Van Sciver has sketched Appleseed lounging on the forest floor, avidly reading a book—again, as with the first image of Appleseed from 1862, an indication of the centrality of Swedenborg's texts for Appleseed's nature-mysticism. This most recent literary instantiation of Appleseed presents his vision as "part of an appeal to what Lincoln called the 'better angels' of an American destiny," as Buhle puts in the introduction: "Chapman's imagined society would foreswear violence in any form, look upon all people as equals but also the animal kingdom as fellow creatures deserving kindness and dignity. It would be far less materialistic, and its members would wish to meet nature more as wanderers than conquerors."[1]

The vestiges of Appleseed's ecological presence in the Midwest also provide the crucial backdrop for key characters in Richard Powers's saga of environmental activism, the much-lauded *The Overstory* (2018). After undergoing a near-death experience where tree-beings appear to have been summoned into her consciousness, Olivia Vandergriff embarks on a quixotic road trip to give her young drifting life more meaning. She unexpectedly finds it in the parking lot of a giant Walmart-like box store in Indiana, while sleeping in her car:

> It's Indiana, 1990. Here, five years is a generation, fifty is archeology, and anything older shades off into legend. And yet, places remember what people forget. The parking lot she sleeps in was once an orchard, its trees planted by a gentle, crazed Swedenborgian who wandered though these parts in rags and a tin pot cap, preaching the New Heaven and extinguishing campfires to keep from killing bugs. A crackpot saint who practiced abstinence while supplying four states with enough fermentable apple mash to keep every pioneer American from nine to ninety half crocked for decades.[2]

Sitting in her bedraggled car, the spirits of Appleseed's trees manifest themselves to Olivia as "beings of light," in a moment of lyrical transfiguration:

> They're everywhere, unbearable beauty, the way they were the night her heart stopped. They pass into and through her body. They don't scold her for forgetting the message they gave her. They simply infuse her again. Her joy at their return spills over, and she starts to cry. They speak no words out loud. Nothing so crude as that. They aren't even *they*. They're part of her, kin in some way that isn't yet clear. Emissaries of creation—things she has seen and known in this world, experiences lost, bits of knowledge ignored, family branches lopped off that she must recover and revive. Dying has given her new eyes.[3]

This crucial moment in the novel suggests how the roots of modern American environmental activism were often grounded in older forms of religious epiphany, a moment of radical self-transformation and conversion—Olivia's vision is the fully mystical, apophatic sort of experience that William James had anatomized in the *Varieties of Religious Experience*, where an encounter with the numinous anchored a noetic truth, felt and cognized in the body. By placing Appleseed's seedlings as these transcendent "beings of light" that lead to Olivia's subsequent involvement in tree-sits and acts of (un)civil disobedience, Powers indicates how Swedenborg's entanglement with American forms of dark green nature spirituality could produce tangible actions, measurable results in the wider public sphere.

A Language of Things has attempted to trace a diverse body of writers and artists who followed in the originary wake of Appleseed and, like him, found within Swedenborg's immanental theology the seeds for their environmental imagination. They did not always transpose Swedenborg's correspondential theories or models of influx into the same keys, but they demonstrate, nevertheless, a consistency of attention to rendering the natural world and the other-than-human in numinous forms, as "vibrant matter" (Jane Bennett), in ways that could counter the prevailing, growing assumption that nature was simply a bunch of random stuff, "out there," waiting to be extracted into the profit-based logic of markets and emergent global capitalism. Ralph Waldo Emerson, John Muir, William Keith, George Inness, and Sarah Orne Jewett all certainly held different views about modernity (and how to make money through their art or writing); and yet, when this book's chapters are taken together, their collective aesthetic engagements with Swedenborg form a consistent pattern around the vitality of what William James would later call "the religious sentiment," demonstrating its rich capacity for filling and shaping environmental imagination.

Environmental historiographies and ecocritical appraisals of the Romantics have tended to subsume their own declensionist narratives about disenchantment through aligning a kind of nonfiction nature-writing with forms of secular humanism; the stubborn persistence of Swedenborg's eighteenth-century theosophy into nineteenth-century eco-aesthetics provides a different set of coordinates for telling the story of how religion could generate alternative environmental imaginaries. This book's primary focus on a disparate set of essays, stories, paintings, and poems further illustrates how the primary greening of Swedenborgian thought took place in non-institutional spaces, outside the confines of the Christian churches that called themselves the New Church, or the New Jerusalem, after Swedenborg's eschatological visions. Some figures, like Inness or Keith, may indeed have been card-carrying Swedenborgians, but they adapted their seer's work on their own terms, and often in heterodox ways (Inness, for example, lived for a period of time in a radical utopian community, and experimented throughout his life with spiritualism—two things largely anathema to the organized New Church).

Even though Swedenborg had aimed for his work to influence the religious institutions of his day, theologically transforming and renovating Christianity from the inside out, by and large his work was either ignored or completely forgotten by later theologians in the nineteenth century (there are some key exceptions here within German-speaking Protestantism, as Friedemann Stengel's important work has recovered).[4] It was rather in the esoteric fringe, and within the domain of the aesthetic, that Swedenborg's ideas came to have their most fruitful encounters, signaled as early as 1809 by William Blake. For the only solo exhibition he put on in his lifetime (held in his brother's hosiery shop), Blake hung sixteen of his most experimental paintings, including an illustration of a visionary scene drawn from Swedenborg's *Apocalypse Revealed*. Blake self-published an accompanying catalogue to this one-man show where he proudly proclaimed that the works of Swedenborg "are well worthy the attention of Painters and Poets," that they were "the Foundations for Grand Things."[5] No one came to Blake's show, but by mid-century, the popularity of Swedenborg had grown so pervasive that Emerson would marvel in his journal that "This Age is Swedenborg's."[6]

A Language of Things has accordingly not attempted any kind of ecclesiastical history of the smaller Swedenborgian New Church movements that often operated in antagonism to Swedenborg's more popular (and aesthetic) diffusions (thus sidestepping the church's propriety claims to be the correct arbiters of Swedenborg's theology). With few exceptions, the Swedenborgians have not been at the vanguard of environmental ecclesiology or other social justice concerns, perhaps due to their tradition's tendency to focus introspectively on individual spiritual regeneration.[7] They have not, for example, followed other strands of Protestantism that, particularly after the wide impact of Pope Francis's encyclical about the Earth, *Laudato Si,* have systematically developed planetary theologies or ethics of Earth care.[8] Nor has this project traced in any systematic way the parareligious and extra-institutional spiritual energies that swept along outside church spaces, forming a heterogeneous "metaphysical tradition" (Catherine Albanese, Courtney Bender) that became the foundations of what we now call (however problematically) the "New Age."[9] Still, the figures in this book—especially Emerson and Muir—have cast clouds of popular influence that extend far outside the domain of literary interpretation and art history: the mind-cure movement and "New Thought" are unthinkable without Emerson, as the progeny of Muir's wilderness spirituality feeds directly into Earth First! and other radical environmental groups in the Pacific Northwest.[10] A basic premise throughout *A Language of Things* has been how forms of environmental imagination became constitutive imaginaries: social practices of collective aesthetic experience that changed and altered American sensibilities about nature and the other-than-human. Reading Emerson's essays, seeing Keith's and Inness's paintings, encountering Muir's *First Summer in the Sierras* or his popular *Stickeen*, or reading Jewett's *A White Heron and Other Stories*, and then joining an Audubon Society—these all formed a habitus of relationality to a numinous nature that had a measurable impact on legal regimes, and interventions made on behalf of conservation or preservation.

In our quickly warming twenty-first century, of what use are such nineteenth-century ecopoetics? There may be certain kinds of Romantic and Transcendentalist baggage best left in the dustbin of history (or better yet: recycled or composted to fit new needs in the Anthropocene—Donna Haraway, for one, talks of our need for more composting instead of more turns toward post-humanisms, the "humusities" of actual humus, rather than the humanities, be they post- or otherwise).[11] Although its aesthetic is clearly still so generative for some kinds of American environmentalism, the wilderness mystique of making "mountains into bibles" increasingly cannot be maintained in a more global and pluralistic form of ecocosmopolitanism that accounts for indigenous cosmologies, especially in the Global South. The old Romantic model of pure, "virgin" nature as a suprahuman space, a wilderness where the human is not, conflicts with indigenous cosmologies that don't bifurcate nature/culture, and see personhood and agency in the other-than-human. And yet, it is still the Romantic sublime that continues to underwrite so much conservation and wilderness management, especially in the so-called third world, where the twentieth-century emergence of international laws around endangered species and habitat preservation has come to reinscribe the paternal order of colonial relations between an enlightened West and benighted Others in the antipodes. On this, *A Language of Things* concurs wholeheartedly with Evan Berry and other historians who have focused on the role of religion in shaping environmental thought; Berry cautions at the end of his *Devoted to Nature* that "environmental advocates ought to be wary of aesthetics as a key value for public deliberations about ecology."[12] The beautiful and the sublime are not transcendent ahistorical values, but shaped and conditioned by the particularity of cultural experience in the West; our topophilia for certain kinds of landscapes over others has been conditioned by various "techniques of the observer" and the related technologies of the panoramic and the picturesque.[13] Viewing landscape as a kind of Protestantized sacred Scripture, rife with correspondences, is by no means a universal mode of perception; Swedenborgian theology easily lent itself to the making of beautiful paintings and the writing of nature essays, but such a language of things needs to be resolutely contextualized and historicized, and not facilely seen as a resolution of a hermeneutic gap between scriptural reading practices and experiences of nature, particularly post-Darwin.

If Swedenborg's particular "science of correspondences" has an environmental future, it would seem to lie in the general ways in which Swedenborgian thought twines around the origin story of (American) biosemiotics: this would have been the book, not yet written, that untangles the Swedenborgian ideas in the background of Charles Saunders Peirce's formation, his schooling in the Swedenborgian household of Henry James Sr., and Peirce's later semiotic theories, which remain important points of departure for this still-growing field.[14] That Jamesian strand of Swedenborgianism provided, as Peirce was to later recall, a "materialism driven deep, and cinched on the inside."[15]

More than the technicalities of Swedenborg's nascent spiritual semiotics, or the epigenesis present in his immanental influx, it is the capacity for wonder enabled by his thought that might offer the most significant legacy for our contemporary crisis. "A force is instilled into everything that has been created," Swedenborg writes in his *Divine Providence:* "However, the force does not do anything on its own; it depends on the one who instilled it. Look at some other subject on our planet. Look at a silk-worm or a bee or some little creature and examine it, first physically, then rationally, and finally spiritually. If you can think deeply, you will be stunned at everything."[16] Wonder as an environmental affect, "stunning the mind," is a drifting mood of enchantment, as Jane Bennett's work has attempted to rehabilitate.[17] It is closely related, as an embodied feeling, to Timothy Morton's more recent articulation of "the Joy" as a way out of the ecological despair and melancholia caused by our powerlessness in the midst of the climate catastrophe—ecological Joy (capital J) "is not abstract or uniform, but so intimate you can't see it, and you can't tell whether its inside or outside: the 'cellular' experience of bonds tightening between beings."[18] But Swedenborg's wonder is also very close, perhaps the other side of the proverbial coin, to another term that has been so equally central for Morton's ecophilosophy: the ecologically weird, or as Morton terms it, really "weird weirdness."[19] Wonder and weirdness stem from the same kind of recognition between ourselves and the animate matter we are part of: Morton's weird weirdness is what happens when wonder loses its teleological grounding in the order of a providential pattern, and untethers from the related ontotheological presumptions of Creation = Creator. In "weird weirdness," it is not the presence of God that stuns the mind, but the uncanny recognition when you turn your car key in the (gasoline) ignition and feel in the thrumming of the motor a connection to species extinction and planetary instability.[20] Ontotheologies like Swedenborg's are precisely what much posthumanist and new materialist thought has gone out of its way to steer clear of; nevertheless, one aspiration of this project has been to keep open a gap between the metaphysics of presence that a mystical figure like Swedenborg provides, and the assorted claims of the new materialisms—a gap perhaps ultimately incommensurable, but one that (I hope) has created certain productive tensions. In their own way, some of the new materialist ontologies with deconstructive roots have ended up reinscribing Enlightenment paradigms of disenchantment and the secular (as in, for example, Quentin Meillassoux's turn toward mathematics for a form of scientific rationalism that can comprehensively account for nonhuman and inorganic realities).[21]

But the "will to believe," as William James called it, seems unlikely to go away any time soon, and *A Language of Things* has aimed to restitute Swedenborg's neglected place in the history of an emergence of an American "nature religion" (Albanese) that increasingly has acquired dark green hues, to paraphrase Bron Taylor.[22] One thing that the religious sentiment, broadly construed, can provide for the Anthropocene (no matter what we call our era of climate crisis) is an alternative set of temporalities that lie outside of, and sometimes chafing against, the teleology of historical progress: that pile of "wreckage upon wreckage" seen by the backward-gazing angel of history in Walter Benjamin's well-known reading of the Paul Klee painting cited at the beginning of this book. Of the more contested claims that Dipesh Chakrabarty has made in his "Climate of History: Four Theses" (2009), which has remained a touchstone for debates about the Anthropocene and its implications for the work of the humanities, is his insistence that, henceforth, species history—its deep time of evolution and our geological force on the planet that marks the Anthropocene—must be "cross-hatched" with the history of capitalism, of modernity and historical consciousness itself.[23] Perhaps in more ways than has tended to be acknowledged, certain kinds of religious structures offer varieties of converging modalities of time that allow for this sort of cross-hatching. As Amitov Ghosh observes, the abysmal failure of effective transnational climate-change legislation points to the insufficiency of the secular nation-state as a building-block (and its attendant ideologies of the bourgeois subject) for climate-change policy: religion contains, contrary-wise, the acceptance of limits and limitations (versus the unfettered growth of free-market capitalism), a homology akin to definitions of the sacred. Religion can be "capable of imagining nonlinear change—catastrophe, in other words—in ways that are perhaps closed to the forms of reason deployed by contemporary nation-states."[24]

In the end, maybe it is Appleseed's unreasonableness, his unsettling "weird weirdness," that has led to his ongoing durability for American environmental imaginaries: the seeming impossibility of this vegetarian pacifist, revering even the mosquitos and the snakes, walking barefoot through the forests while he distributed his Latinized Swedenborg on an illiterate and often-hostile frontier. His folkloric iconicity registers a different kind of time that, however brief, could flair and flash its proto-ecological apple boughs within the violent depravations of colonial modernity. Appleseed, as Buhle writes in the introduction to their historical comic, shows "the paths offered but not taken by society." A later twentieth-century poet like Vachel Lindsay could claim Appleseed as part of a visionary-aesthetic American tradition that harbored anti-Modernist, or countercultural, valences; Appleseed "was the New England kind of saint, much like a Hindu saint, akin to Thoreau and Emerson who came after him."[25] Lindsay was very "glad Johnny Appleseed was a Swedenborgian," he wrote in 1920, because "a good Swedenborgian is a better enemy of economic tyranny than any economist."[26] A deep reader of Swedenborg himself, in Lindsay's later long poem about Chapman, "In Praise of Johnny Appleseed," he amplifies the unusual temporalities that Chapman's legend suggests, conflating an American past with a mystical future-to-come:

> Then
> "Johnny Appleseed, Johnny Appleseed,"
> Chief of the fastnesses, dappled and vast,
> In a pack on his back,
> In a deer-hide sack,
> The beautiful orchards of the past,
> The ghosts of all the forests and the groves—
> In that pack on his back,
> In that talisman sack,
> To-morrow's peaches, pears, and cherries,
> To-morrow's grapes and red raspberries,
> Seeds and tree-souls, precious things,
> Feathered with microscopic wings . . .[27]

In one sense, Lindsay's rhythmic lines about tree-souls have become literally true: along the Ohio River Valley one can still find apple trees that are reputed to descend from Appleseed's original plantings, flowering and bearing their sour cider apples, two centuries later. Appleseed's ecological tomorrow ghosts our today. And perhaps somewhere out there in the weirder corners of Americana, the seeds of Swedenborg's immanental theology are continuing to cross-hatch, maybe to sprout still stranger forms of eco-aesthetics for the Anthropocene: harbingers of an American nature that seems likely to remain, for many, numinous and vibrant.

Notes

Abbreviations

AC	Emanuel Swedenborg, *Arcana Caelestia*
BLUC	Bancroft Library Special Collections, University of California, Berkeley
CEC	Ralph Waldo Emerson, *Correspondence of Emerson and Carlyle*
CL	Emanuel Swedenborg, *Conjugial Love*
CSS	Archives of the Center for Swedenborgian Studies, Berkeley, California
DD	Emanuel Swedenborg, *Swedenborg's Dream Diary*
DLW	Emanuel Swedenborg, *Divine Love and Wisdom*
DP	Emanuel Swedenborg, *Divine Providence*
EL	Ralph Waldo Emerson, *Essays and Lectures*
EP	Ralph Waldo Emerson, *Essays and Poems*
EU	Emanuel Swedenborg, *Earthlike Bodies Called Planets*
HH	Emanuel Swedenborg, *Heaven and Hell*
IBSB	Emanuel Swedenborg, *Interaction between Soul and Body*
JMN	Ralph Waldo Emerson, *Journals and Miscellaneous Notebooks*
KVA	Archives of the Kungliga Vetenskapsakadamien (Royal Swedish Academy of Sciences), Stockholm
LMS	Emanuel Swedenborg, *Letters and Memorials*
LRE	Ralph Waldo Emerson, *The Letters of Ralph Waldo Emerson*
PR	Alfred North Whitehead, *Process and Reality*
SE	Emanuel Swedenborg, *Emanuel Swedenborg's Diary of Spiritual Experiences*
SFSC	Archives of the San Francisco Swedenborgian Church
SR	Sampson Reed, *Sampson Reed: Primary Source Material for Emerson Studies*
SSL	Archives of the Swedenborg Society, London
TCR	Emanuel Swedenborg, *True Christianity*
UoP	Holt-Atherton Special Collections, Archives of the University of the Pacific
WJW1	*William James Writings, 1878–1899*
WJW2	*William James Writings, 1902–1910*

Introduction

1. Benjamin, *Illuminations*, 258.

2. In spite of Latour's hesitation to invoke the "tired" trope of Benjamin's angel, he offers a fresh and innovative reading of the rubble of progress as generated by the angel itself, and modernity's (puritanical) paranoia to separate itself from the past: "The ecological crisis is nothing but the sudden turning around of someone who had actually never before looked into the future, so busy was He extricating Himself from a horrible past." Latour, "Compositionist Manifesto," 486.

3. Davenport and Robertson, "Resettling the First American 'Climate Refugees'"; as these Louisiana islanders flee to the U.S. mainland, other scientists have documented how five of the Solomon Islands have already disappeared due to rising ocean levels caused by climate change. Albert et al., "Interactions between Sea-Level Rise and Wave Exposure."

4. DeConto and Pollard, "Contribution of Antarctica," passim; apart from the challenge this presents to human life, DeConto and Pollard further note how "the long thermal memory of the ocean . . . will inhibit the recovery of marine-based life for thousands of years after greenhouse gas emissions are curtailed" (which assumes that such a global curtailing is even possible), 596.

5. IPCC, "Global Warming of 1.5° Celsius."

6. Mooney, "Earth's Atmosphere Just Crossed."

7. Wallace-Wells, *Uninhabitable Earth*, 34.

8. Klein, *This Changes Everything*, 12.

9. Kolbert, *Sixth Extinction*.

10. Haraway, *Staying with the Trouble*, 30; as Haraway emphasizes, quoting another thinker she dialogues with (Marilyn Strathern): "[I]t matters what ideas we use to think other ideas."

11. White, "Historical Roots of Our Ecological Crisis," 1206. As Bron Taylor notes, "White thought that although Western people were becoming increasingly secular, the West's 'set of basic values' are nevertheless shaped by Christianity, and therefore, 'we shall continue to have a worsening ecologic crisis until we reject the Christian axiom that nature has no reason for existence save to serve man.'" B. Taylor, "Greening of the Religious Hypothesis," 271.

12. Latour, *We Have Never Been Modern*.

13. Weber, "Science as a Vocation," in *From Max Weber*, 140.

14. See the discussion of Swedenborg and the Swedenborgians in Albanese, *Nature Religion in America*, 110–34. See also B. Taylor's discussion of Emerson, Muir et al., as foundational religious figures in *Dark Green Religion*, 42–54.

15. On Watt and Swedenborg, see R. K. Silver, "Spiritual Kingdom in America," 307. For how Swedenborg's work on mechanics helped catalyze the Industrial Revolution in Sweden, see Lindqvist, *Technology on Trial*, 158–70; Jenny Uglow has more recently noted, in passing, the utility of Swedenborg's mineralogy for the Lunar Men Society in London in *Lunar Men*, 282. Watt's steam engine is cited in the debut of the

"Anthropocene" as a geological concept, in Crutzen and Stoermer's now-landmark "The Anthropocene," 17–18.

16. The "fact of feeling" turn of phrase is indebted to Joan Richardson, whose work on Swedenborg in American literature has been foundational for this present book. Richardson, *Natural History of Pragmatism*, esp. 7–8.

17. Castoriadis, *Imaginary Institution;* C. Taylor, *Modern Social Imaginaries.*

18. Kerrigan, *Johnny Appleseed*, 110; Knapp, *History*, 32–33.

19. Price, *Johnny Appleseed*, 139.

20. Pollan, *Botany of Desire*, 35.

21. P. Miller, ed., *The Transcendentalists*, 49.

22. Buell, *Environmental Imagination*, esp. chs. 1 and 2; and Nash, *Wilderness*, 84–95.

23. Abrams, *Natural Supernaturalism*, 12; Gatta, *Making Nature Sacred*, 12; Rigby, *Topographies of the Sacred*, 12, 45–46.

24. On Leibniz and Spinoza, see Skrbina, "Panpsychism in History," 11–14; and *Panpsychism and the West*, 87–90, 95–100; for a similar turn to Schelling, see Grant, *Philosophies of Nature*. Bennett's *Vibrant Matter* makes substantial recourse to Spinoza.

25. Stengers, *Thinking with Whitehead.*

26. Pierce, *Collected Papers of CSP*, vol. 8, par. 55–90.

27. Deleuze, *Logic of Sense*, 311.

28. Haller, *Swedenborg;* Gabay, *Covert Enlightenment.*

29. Emerson, *JMN*, 12:335–36.

30. Matthiessen, *American Renaissance*, viii.

31. Schmidt, *Hearing Things;* Albanese, *Nature Religion in America* and *Reconsidering Nature Religion.*

32. On the December 20, 2011, *Jeopardy!* show, syndicated by NBC, under the category "Touched by an Angel," Swedenborg was the $1,600 answer to "this eighteenth-century Swedish scientist who had visions of angels gave his name to a Christian mystic movement." As my friend T. Lester brought to my attention, Swedenborg's summer house appeared more recently on *Jeopardy!* (January 15, 2019) in a question under the thematic "A Visit to Skansen" (the open-air museum in Stockholm where Swedenborg's *lusthus* presently resides).

33. Suzuki, *Swedenborg;* Zuber, "Buddha of the North."

34. Gaard, "Mindful New Materialisms."

35. C. Wolfe, *What Is Posthumanism?*, xx; Bergthaler, "Limits of Agency," 38; Meillassoux, *After Finitude*, 45–47.

36. My emphasis; Reed, *SR*, 33.

37. Buell, *Environmental Imagination;* Nash, *Wilderness.*

38. Emerson, *EP*, 21.

39. Emerson, *EP*, 22.

40. Walls, *Passage to the Cosmos;* Sachs, *Humboldt Current;* Wulf, *Invention of Nature.*

41. The Lacey Act made the interstate transport of illegally killed wildlife a federal offense. It was specifically fomented by George Grinnell and other ornithologists associated with the Audubon Society. See Nash, *Rights of Nature*, 49–50.

42. Otto, *Idea of the Holy,* 12–23, 31.

43. Schneck, "'The Great Secular Transcendence,'" 214.

44. Rose, "Shimmer," G53–55; Bennett, *Vibrant Matter.*

45. See Hallowell, "Ojibwa Ontology," 141–79; and more recently, Harvey, *Animism,* passim.

46. Bergquist, *Biblioteket i lusthuset,* 18–20.

47. Swedenborg, *DLW,* no. 56. Following standard practice in scholarship on Swedenborg, this and all subsequent citations of his work refer to section (and not page) numbers.

48. Swedenborg, *DLW,* nos. 332, 3. On Swedenborg and Linnaeus, see Bergquist, "Linnaeus and Swedenborg," 23–39.

49. J. J. Cohen, "Foreword," x.

50. Bennett, *Vibrant Matter,* 13.

51. Swedenborg, *TCR,* no. 80.

52. Butterfield, "The Genius of George Inness," 9.

53. Borges, *Selected Non-Fictions,* 452.

54. "This inner sense is unknown to human beings, and it is this that Schwedenberg [*sic*], whose inmost soul is opened up, wishes to make known to man." Kant, *Kant on Swedenborg,* 53. As Wouter Hanegraaff puts it, "Der Hauptpunkt des erstens Kapitals [von *Traümes*] ist wesentlich für das Buch als Ganzes: Kant argumentiert, dass es zwar nicht irrational sei, die theoteritische *Möglichkeit* immaterialler Wesen anzunehmen, dass es aber unmöglich sei, die tatsächliche *Existenz* solcher Wesen entweder zu beweisen oder zu widerlegen." Hanegraaff, "Swedenborg aus der Sicht von Kant," 158.

55. Gregory Johnson argues that "Swedenborg's descriptions of the spirit world and its pneumatic law in the *Arcana Coelestia* are the models of Kant's accounts of the moral world as a 'kingdom of ends' [*Reich der Zwecke*] in the *Groundwork* and of the afterlife in the *Critique of Practical Reason*. . . . Swedenborg actually refers to the spiritual world as a 'kingdom of ends' (*regnum finium*) in the *Arcana Coelestia,* which is the earliest use of the phrase I can ascertain." Johnson, "Introduction," xviii. As a further point of clarification regarding some of the variant spellings here, Swedenborg's original *Arcana Coelestia* volumes were spelled with the Latin "œ"; English translations in the eighteenth century began using the variant "æ," so that "Cœlestia" became "Cælestia." Starting in the nineteenth century and continuing into the present, the "æ" was often then modernized to an "ae," and "Cœlestia" became "Caelestia." This is important, as unlike all of Swedenborg's other titles that were readily translated into English, the Latinate *Arcana Caelestia* was usually retained for this *summa theologica* (albeit in a modern spelling that would have seemed odd to Swedenborg).

56. These famous lines are from William Blake's manuscript poem now called "The Auguries of Innocence." Blake, *Complete Poetry and Prose,* 490–92.

57. On the dilemmas of what to name our era—Anthropocene, Chthulucene, Capitalocene, etc.—see Haraway, *Staying with the Trouble,* 44–57.

58. This is how Nils Bubant phrases it; "The Anthropocene . . . invites us to imagine a world in which an alien geologist from the future detects in the strata of the

ground evidence of the presence of humans long after we have gone extinct." Bubandt, "Haunted Geologies," G135.

59. Francis Fukayama iconically read the demise of Soviet-era communism as the "end of history," as the termination of a struggle of competing ideological worldviews and the definitive triumph of global capitalism and (neo)liberal institutions (Fukayama, *End of History*). This viewpoint was severely qualified by 9/11 and its geopolitical aftermath, and, more recently, by the unfolding climate crisis; post-Trump, Fukayama has tempered his earlier triumphalism about democracy: Fukayama, "Is American Democracy Strong Enough for Trump?" On climate change and the crisis facing historiography, see Chakrabarty, "Climate of History," 197–222.

60. See, for example, the surviving letters that attest to Swedenborg's efforts to acquire (then) exotic vegetables like cucumber and eggplant, and ensuing confusion on the part of seed and bulb distributors who had never heard of the strange plant names. Swedenborg, *LMS*, 511–13.

61. Swedenborg, *HH*, no. 489. Emerson copied parts of this passage from Swedenborg into Journal Z, some year prior to 1838. He sounds, however, a note of irritation at Swedenborg's detailed particularity: "But it is vicious to say . . . of the fair shrubs & trees which adorn the gardens of the angels, that 'these are changed every day,' <like a city lady's row of plants.> as if he was thinking of <a shop window of grand shop> the prints in [a] city bookseller's shopwindow." *JMN*, 6:315.

62. Emerson, *EP*, 29.

63. Reynolds, *Walt Whitman*, 94–96; Hallengren, "Hermeneutic Key," 45–59.

64. Traubel, *With Walt Whitman*, 5:376.

65. As noted by Harding and Meyer, *New Thoreau Handbook*, 98.

66. Harding, "Swedenborgian Spirit"; Richardson, *Thoreau*. Laura Dassow Walls in both her meticulously researched *Seeing New Worlds* and *Henry David Thoreau* gives nary a word to Swedenborg.

67. Buell, *Environmental Imagination*.

68. Poe, *Eureka*, 36, 135, 140.

69. Ludwig Wittgenstein considered language-making more like a game than the painting of pictures—a "Sprachspiel"—in his *Philosophical Investigations*, of course, but the playfulness of semiosis is also essential for American Pragmatism; see Richard Poirier on the "saving uncertainty and vagueness" of American poetics, a "pragmatist poetic dispensation" that begins with Emerson. *Poetry and Pragmatism*, 4, 80.

70. James, *WJW2*, 509.

1. Planetary Pictures

1. Jonsson, *Visionary Scientist*, 63.

2. Swedenborg, *Principia*, 15.

3. This is according to Odhner, *Wilkinson*, 19.

4. See Peet, *Medicine, Mysticism, and Mythology*, 10–11, 84–91.

5. J. J. G. Wilkinson, *Swedenborg*, 69.

6. Walls, *Passage to the Cosmos*, 251–52.

7. Bergthaler, "Limits of Agency," 38; Meillassoux, *After Finitude;* Morton, "X-Ray," 315.

8. One of most significant theological inheritors of Whitehead remains Catherine Keller; see her *Cloud of the Impossible* for an environmental inflection of process thought toward "a theopoetics of nonseparable difference," 11, and chapter 5 for a reading of Whitehead through the Deleuzian fold.

9. Barad, *Meeting the Universe*, 356.

10. Whitman, *Portable Walt Whitman*, 8.

11. Harman, *Third Table*, 12, 15.

12. Bergquist, *Swedenborg's Secret*, 282.

13. Swedenborg, *DP*, nos. 190, 202.

14. Whitehead, *PR*, 351.

15. Bergquist, *Swedenborg's Secret*, 100–101.

16. Dunér, *Natural Philosophy*, 80.

17. Jonsson, *Visionary Scientist*, 22.

18. Swedenborg, *Daedalus* 5 (1716): 132.

19. "Swedenborg's doctrine of the senses is an 'iatrogeometry' in which vision extends towards the other senses. All sensory impressions could be explained by the mechanistic natural philosophers in terms of geometry and mechanics; they could be returned to the fundamental primary properties. . . . Swedenborg's iatrogeometry is a Cartesian reduction of the sense to extent and movement." Dunér, *Natural Philosophy*, 206.

20. Qtd. in Dunér, *Natural Philosophy*, 140.

21. Qtd. in ibid., 181.

22. Fors, *Limits of Matter*, 124.

23. Dumas and Bineau, *Leçons sur la philosophie chimique*, 215.

24. Haraway, *Crystals*, 41.

25. The *Prodromus Principorium* was later translated into English as *Some Specimens.*

26. See, for example, the argument that Swedenborg suffered from temporal lobe epilepsy, Foote-Smith and Smith, "Emanuel Swedenborg," 211–18.

27. Foucault, *Madness and Civilization*, especially chap. 7, 199–220; Foucault, *Archeology of Knowledge.*

28. Swedenborg, *LMS*, 696.

29. Fors, *Limits of Matter*, 142.

30. Benz, *Swedenborg*, 108–9.

31. Tafel, *Swedberg*, passim; Bergquivst, *Swedenborg's Secret*, 7–14.

32. Berlin, *Roots of Romanticism*, 49.

33. Swedenborg, *SE*, no. 3320.

34. Swedenborg, *DD*, 175.

35. See Hallengren, *Gallery of Mirrors*, 23–41.

36. As noted by Dole and Kirven, *Scientist*, 4, 90.

37. Swedenborg, *SE*, no. 419; Swedenborg, *Ludus Heliconus*, 111.

38. Robsahm, *Memoirs*, 4, 25.

39. Swedenborg, *Interaction*, no. 11.

40. Swedenborg, *Ludus Heliconus*, 111.

41. In the poem "A description of the most distant lapps," Swedenborg uses the word "wild" repeatedly: "being wild themselves, they drive wild animals into their nets." The Neolatin that Swedenborg uses here is "ferus"—from which our modern English "feral" descends.

42. J. J. G. Wilkinson, *Swedenborg*, 32.

43. Schuchard sensationally claims that Swedenborg practiced tantric sexual yoga and sexual kabbalistic rites, which brought on "orgasmic trances" that elevated Swedenborg into the realm of angels—see *Secret Agent*, 367, 442; also Schuchard, "Why Mrs. Blake Cried," 55. These erotic commitments were apparently in addition to Swedenborg's previously unknown career as a secret agent working for various political causes. Schuchard's large biography of Swedenborg reads better as espionage fiction than as reliable scholarship; some of her many errors and distortions are catalogued by Stengel, "Rezension über Marsha Keith Schuchard," 113–29.

44. Hallengren, *Grand Theme*, 90.

45. Swedenborg, *Philosopher's Notebook*, 75.

46. Emerson, *EP*, 663.

47. Deleuze, *Difference and Repetition*, 237.

48. Swedenborg, *HH*, no. 106.

49. Lindqvist, *Technology on Trial*, 158–70; R. K. Silver, "Spiritual Kingdom," 307; Uglow, *Lunar Men*, 282.

50. Swedenborg, *Principia*, 2:255.

51. Whitehead, *Science*, 10.

52. Dunér, *Natural Philosophy*, 340.

53. Swedenborg, *Principia*, 2:263.

54. Swedenborg first references Fludd during his initial sojourn to England; see Swedenborg, *LMS*, 1:55.

55. Rosenberg, "Was There a First Abstract Painter?" 91–93.

56. "Multiverse," while used today to denote various hypothetical scenarios within quantum physics and string theory, was a term originally coined by William James. See Rubenstein, *Worlds without End*, 3–5; Rubenstein also very briefly acknowledges Swedenborg's nebular hypothesis, 275.

57. Dunér, *Natural Philosophy*, 346.

58. Lindroth, *Swedish Men of Science*, 55.

59. Latour, "Compositionist Manifesto," 480.

60. Rubenstein, *Worlds without End*, 18.

61. Whitehead, *PR*, 94–95.

62. My emphasis; Swedenborg, *Philosopher's Notebook*, 244.

63. Bergquist, *Swedenborg's Secret*, 116.

64. Swedenborg, *Principia*, 2:375.

65. Ek, Löfgren, and Bergholm, "Environmental Effects," 96–103.

66. Bubandt, "Haunted Geologies," G122.

67. Ibid., G135–36. More recently, Bubandt has expanded his call for broader nonsecular approaches to the Anthropocene, noting the "absence of sustained, empirical exploration of the ephemeral, spiritual, magical qualities of the nonhuman agency that has come to take center stage." Bubandt, "Anthropocene Uncanny," 7.

68. Fors, *Limits of Matter*, 144.

69. Swedenborg, *Principia*, 2:372–73.

70. Derrida, *Specters of Marx*, 9–10.

71. Rigby, *Topographies of the Sacred*, 158.

72. This is in terms of how Dr. Frankenstein refuses to create a mate for the monster in chapter 20 of the novel, out of the fear that the two creatures would subsequently procreate a (superior) "race of devils" who "might make the very existence of the species of man a condition precarious and full of terror." Dr. Frankenstein's fear about unleashing a kind of holocaust—"had I right, for my own benefit, to inflict this curse upon everlasting generations?"—has been read as a keenly ecological sort of predicament (and not only the iconic precursor for a whole genre of dystopian futurism). Shelley, *Frankenstein*, 173–74; for ecocritical appraisals of Shelley's iconic text, see Morton, "Frankenstein and Ecocriticism," 143–57, and Carroll, "Crusades," 211–30.

73. Emphasis in the original Latin; Swedenborg, *Ludus Heliconus*, 124–25.

74. This occurs in the gospel of John, chapter 19. Swedenborg relied on a number of Latin versions of the Bible, including Van der Hooght and Schmidius's *Biblia Hebraica* (1740) and Schmidt's *Biblia Sacra* (1696). His heavily annotated copies are in the Royal Swedish Academy of Sciences archives—see KVA, Codexes 64, 73, 89, 90. Partial English translations of Swedenborg's Bible marginalia are available as Swedenborg, *Schmidius Marginalia*.

75. Iovino, "Behold," 312, 324.

76. Bubandt, "Haunted Geologies," G124–25.

77. There is little about nonhuman animals in either Swedenborg's *Animal Kingdom* or *Economy of the Animal Kingdom*, as we shall soon see. There remains chronic confusion about these (mis)translated titles of Swedenborg's original *Oeconomia Regni Animales* and *Regnum Animale*. As Charles Gross points out, a better translation for the *Oeconomia* would be something like "Biological bases of the soul"; *Brain*, 123. However, given that Emerson and other figures in this book would have known Swedenborg's physiological work as the *Animal Kingdom* and the *Economy* from the nineteenth-century translations undertaken by Clissold and Wilkinson—which have yet to be superseded by any more modern editions—to avoid further confusion, *A Language of Things* sticks with the earlier misnomers of "animal" in the titles.

78. Gross, *Brain*, 127.

79. Finger, *Origins*, 29–31.

80. Lakoff and Johnson, *Metaphors*.

81. Jonsson, *Visionary Scientist*, 38, 70; Bergquist, *Swedenborg's Secret*, 134–35.

82. Swedenborg, *Animal Kingdom*, 1:451.

83. Instead of simple location, "the body is the organism whose states regulate our cognizance of the world. The unity of the perceptual field must be a unity of bodily

experience. . . . [T]his involves the entire abandonment of the notion that simple location is the primary way in which things are involved in space-time." Whitehead, *Science*, 91.

84. Swedenborg, *DD*, 92.

85. Swedenborg, *Worship*, no. 66.

86. Emerson, *EP*, 49.

87. Dunér, *Natural Philosophy*, 364.

88. Horn, *Schelling und Swedenborg;* for a counterargument, that Krause was actually much more critical and distanced from Swedenborg, see Forster, *Karl Christian Friedrich Krause*. See also Stengel, *Aufklärung*, 404.

89. Rigby, *Topographies of the Sacred*, 48.

90. Morton, *Dark Ecology*, 19.

91. Dr. Gabriel Beyer (a Swedish adherent of Swedenborg's based in Gothenburg) noted the profound parallels between Swedenborg and Böhme, and asked Swedenborg if he was familiar with the earlier German mystic. Swedenborg responded in the negative, perhaps disingenuously: "I have never read [the books of Böhme], and it was forbidden me to read dogmatic and systematic books in theology before heaven was opened to me, and this for the reason that thereby unfounded opinions and novelties might find occasion to insinuate themselves, which afterwards could be removed with difficulty. Therefore, when heaven was opened to me, I had to first learn the Hebrew language, and also correspondences, of which the whole Bible is composed, and this led me reread the whole of God's Word many times." Swedenborg, *LMS*, 2:630.

92. Bennett, *Enchantment of Modern Life*, 33–40.

93. Swedenborg, *Worship*, no. 10.

94. Swedenborg, *Worship*, nos. 32–33.

95. "But there was still wanting a being who could refer these pleasures of the senses to a sort of proper mind [*mentem*], or to his own consciousness and perception, and who, from the faculty of intellect, might decide upon the beauty resulting from all these harmonies, and might from beauties perceive joys . . . there was wanting, I say, that son of earth, or that mind under a human form, which from the paradise of earth might look into the paradise of heaven." Swedenborg, *Worship*, no. 29.

96. Swedenborg, *Worship*, nos. 94, 43, 75.

97. Jonsson, *Drama of Creation*, 16.

98. Swedenborg, *Worship*, no. 33.

99. Jonsson, *Drama of Creation*, 103, 108.

100. Deleuze and Guattari, *Thousand Plateaus*, 249.

2. Psychogeographies of Heaven and Hell

1. The phrase "heaven, hell, and other places" is cribbed from Gary Lachman—the former drummer of Blondie, now turned estoricist—in the 2011 eponymous documentary about Swedenborg, *Heaven, Hell, and Other Places*, directed by Jacob Cartwright, Nick Jordan, and Stephen McNeilly.

2. Borges, *Selected Non-Fictions,* 457.

3. Miłosz, *ABCs,* 245; Borges, *Selected Non-Fictions,* 452. For Borges's profound and sustained response to Swedenborg, see Rowlandson, *Borges,* 193–224.

4. Swedenborg, *HH,* no. 38.

5. See Swedenborg, *SE,* no. 5471; KVA, Codices 2, 3A, and 3B.

6. Psychogeography is "the study of the precise laws and specific effects of the geographical environment, consciously organized or not, on the emotions and behaviors of individuals," though it quickly acquired a political valence as a means for working against the commodity fetishism of capitalism. Debord, "Introduction." For a gloss on reading Swedenborg via Debord, in context of William Blake and what Iain Sinclair calls Blake's "topographic sublime," see Zuber, "Apocalypse of Mind."

7. Baudelaire's "Correspondance" is central to Benjamin's *Arcades* project; see 241, 248, 292–98. See also Zuber, "Flânerie," passim.

8. Sinclair, *Blake's London,* passim; and Sinclair, *Swimming to Heaven.*

9. On Blake's urban ecopoetics, see Hutchings, *Imagining Nature,* 60–66; and Zuber, "Satanic Mills," 39–49.

10. Dunér, *Natural Philosophy,* 416.

11. Stengel, *Aufklärung,* 324–34.

12. Swedenborg, *DD,* 304.

13. Swedenborg, *SE,* no. 2060.

14. Suzuki, *Swedenborg;* Zuber, "Buddha of the North"; Tweed, "Occultism."

15. Miłosz, *ABCs,* 274.

16. Gaard, "Mindful New Materialisms," 291–300; Rigby, "Spirits That Matter," 283–90.

17. Hanegraaff, "Swedenborg, the Jews, and Jewish Traditions"; Dole, "Philosemitism"; Esterson, "Secrets of Heaven." Most recently, James Lawrence has convincingly shown parallels between Swedenborg and the kabbalistic traditions thorough gendering of every aspect of their respective metaphysics. This is so extensive that "the gendered aspect" of spiritual correspondence becomes the basis of biblical interpretation: "Creation itself is thus a profoundly genderized unfolding of the divine into manifestation such that all things in creation are irreducibly gendered as a fundamental feature of the divine." Lawrence, "Correspondentia," 59.

18. See Deleuze and Guattari on the "refrain" in *Thousand Plateaus,* 310–50; Bennett on chanting and enchantment in *Enchantment of Modern Life,* 6, 36; Abram on spells and spelling, and the alphabet, in *Spell,* 100, 133.

19. "Can this becoming, this emergence, be called Art? That would make the territory a result of art. The artist: the first person to set out a boundary stone, or to make a mark. Property, collective or individual, is derived from that even when it is in the service of war and oppression. Property is fundamentally artistic because art is fundamentally *poster, placard.*" Deleuze and Guattari, *Thousand Plateaus,* 316.

20. My emphasis; qtd. in Bergquist, *Swedenborg's Secret,* 157.

21. Richardson, *Natural History of Pragmatism,* 169.

22. Hanegraaff, *New Age Religion,* 428.

23. Derrida, *Of Grammatology;* Derrida protested that he never "ceased to be surprised by critics who see my work as a declaration that there is nothing beyond language, that we are imprisoned in language; it is, in fact, saying the exact opposite. The critique of logocentrism is above all else the search for the 'other' and the 'other of language.'" Kearny, *Dialogues with Contemporary Thinkers,* 123.

24. Wheeler, *Whole Creature,* 120–23; Wheeler, "Natural Play," 67–79.

25. Peirce, "Basis of Pragmaticism," 394.

26. Swedenborg, *HH,* nos. 234–45.

27. A. Taylor, *Henry James,* 8–21; Orange, *Peirce's Conception of God,* 5; Croce, *Science and Religion,* 56–60. See also E. Taylor, "Peirce and Swedenborg."

28. Blumenberg, *Lesbarkeit der Welt,* especially 267–81.

29. "Philosophy is explanatory of abstraction and not of concreteness. It is by reason of their instinctive grasp of this ultimate truth, that, in spite of much association with arbitrary fancifulness and atavistic mysticism, types of Platonic philosophy retain their abiding appeal; *they seek the forms in the facts.*" Whitehead, *PR,* 27.

30. See L. Wilkinson, *Dream,* 239–47.

31. Baudelaire, *Selected Poems.*

32. Qtd. in Herzogenrath, "White," 13–14.

33. Herzogenrath, "'Weather of Music,'" 227.

34. Swedenborg, *AC,* no. 2853.

35. See, for example, Swedenborg, *AC,* no. 5391.

36. Emerson, *EP,* 1040.

37. Deleuze and Guattari, *Thousand Plateaus,* 40.

38. Swedenborg, *EU,* no. 132.

39. Swedenborg, *EU,* no. 43.

40. Hallengren, *Grand Theme,* 47.

41. Swedenborg, *EU,* no. 13.

42. Deleuze and Guattari, *Thousand Plateaus,* 154.

43. Ibid.

44. Ibid., 158–59.

45. Ibid., 160.

46. Rauer, *Wahn und Wahrheit,* 206, 233.

47. Stengel, *Aufklärung,* 638–640.

48. Swedenborg, *EU,* no. 49.

49. Swedenborg, *EU,* no. 173.

50. Swedenborg, *SE,* no. 3867.

51. Swedenborg, *EU,* no. 55.

52. Swedenborg, *EU,* no. 112.

53. Holmes, *Age of Wonder,* 122–123.

54. Heise, *Sense of Place,* 22–26; Dunaway, *Natural Visions,* 207–9.

55. Milton, *Paradise Lost,* 171.

56. Swedenborg, *EU,* no. 126.

57. Swedenborg, *DLW,* no. 344.

58. Swedenborg, *DLW,* nos. 311, 314.

59. Swedenborg, *DLW,* no. 73.

60. "Trans-corporeality" is Stacy Alaimo's well-known turn of phrase, clarified at length in her *Bodily Natures;* it is "a new materialist or posthumanist sense of the human as perpetually interconnected with the flows of substances and the agencies of environments. Activists . . . work to reveal and reshape the flows of material agencies across regions, environments, animal bodies, and human bodies—even as global capitalism and the medical-industrial complex reassert a more convenient ideology of solidly bounded, individual consumers and benign, discrete, products." Alaimo, "Oceanic Origins," 187.

61. Gigante, *Life,* 9.

62. Swedenborg, *DLW,* no. 432; Johnson, *Visionary Scientist,* 63.

63. Swedenborg, *DLW,* no. 432.

64. Richards, *Romantic Conception,* 12.

65. Emerson, *EP,* 5.

66. Emerson, *EP,* 10–11.

3. Radical Correspondence

1. Swedenborg, *Theosophic Treatise,* no. 19. In his copious commonplace book of philosophical quotes, aphorisms, and reflections—published only posthumously—Swedenborg seems keenly attuned to the convergence between these different positions, in spite of the ways they appear clearly demarcated in the later theology. He notes, approvingly: "In his Letters, Leibnitz says respecting Syncretism: I find that the opinions of Malebranche are not far removed from my own. The transition from Occasional Causes to Pre-established Harmony does not appear to be overly difficult. In any case, if any one wishes to take the System of Occasional Causes in a way which transforms it into my position I will not object." Swedenborg could be said to have undertaken such a transformation through his subsequent panentheistic theosophy. Swedenborg, *Philosopher's Notebook,* 353–54. On labeling Swedenborg's theology as panentheistic, see prior discussion on pages 51–52.

2. Emerson, *EP,* 10.

3. "In short, the spirit and peculiarity of that impression nature makes on us, is this, that it does not exist to any one or to any number of particular ends, but to numberless and endless benefit; that there is in it no private will, no rebel leaf or limb, but the whole is oppressed by one superincumbent tendency, obeys that redundancy, or excess of life which in conscious beings we call *ecstasy.*" Emerson, *EP,* 121. On Emerson's relation to Darwin, see Walls, *Emerson's Life in Science,* 167–68; and Richardson, *Natural History of Pragmatism,* 88.

4. Emerson, *EP,* 20.

5. Cameron, *Emerson,* 1:302.

6. Emerson, *EP,* 688.

7. James McKusick (*Green Writing*), Andrew McMurray (*Environmental Renaissance*), and John Gatta (*Making Nature Sacred*) are among those who have worked

to rehabilitate Emerson's reputation for ecocriticism; more recently, *"Best Read Naturalist,"* an edited collection of Emerson's writings on nature, promises to further restore Emerson as a "green" thinker and writer. Nicholas Friesner also critically works against the secessionist narrative of Emerson as stepping-stone to the more earthy (better) Thoreau by arguing Emerson's Idealism might have something profound to offer for the Anthropocene; Friesner, "Transcendentalist," 1–18. McKusick's juxtaposition of European Romantic currents with the development of an American environmental ethos comes closest to this book's preoccupations; while I wholeheartedly agree with McKusick's contention that American literature has long suppressed its debt to European Romanticism through a "self-begotten" myth of wilderness encounters, I believe it is more Swedenborg, less Samuel Taylor Coleridge, that lies behind Emerson's specific views of nature as "correspondence," and as sacral language. *Green Writing*, 127.

8. Brøgger, "Conception of Nature," 96–97.

9. The neglected mystical, ecstatic, and religious dimensions of Thoreau have been artfully recovered by Kevin Dann in his *Expect Great Things*, especially chapter 2, "Seeing the Unseen," 40–95.

10. Emerson, *EP*, 11.

11. See the collection, edited by Skrbina, *Mind That Abides;* and Matthews, *For Love of Matter.*

12. See Harding, *Emerson's Library;* and Hotson, "Emerson's Biographical Sources," 23–46.

13. Richardson, *Emerson*, 100–101.

14. For the central importance of the lecture circuit for shaping Emerson's thoughts, see Voelz, *Transcendental Resistance.*

15. Rorty, *Mirror*, 192–93; Richardson, *Natural History of Pragmatism*, 257–58n43.

16. Emerson, "Inspiration" in *The Complete Works*, 292.

17. Emerson, *EP*, 256.

18. Emerson, *EP*, 266.

19. Emerson, *EP*, 457.

20. "Ecopoetics" gained currency in ecocriticism following the group of poets and critics who published in *ecopoetics*, during that eponymous journal's run under Jonathan Skinner (which ceased publication in 2009). *Ecopoetic*'s critical genealogy is indebted less to the nonfiction prose genre inaugurated by Buell's *Environmental Imagination*, and more to the ecocritical appraisals of Romantic poetry that have argued, often via strategic deployments of Martin Heidegger, that poetic language occupies a particularly (privileged) space within environmental relations—see Fletcher's *New Theory for American Poetry* and Bate's *Song of the Earth*, though neither uses the term "ecopoetics" per se. Since Skinner's influential journal, "ecopoetics" has been employed in important collections of ecocritical theory, poetry, and poetics, including essays and work in Iijima's *Eco Language Reader* and Corey and Waldrep's *Arcadia Project*. See also Skinner, "What Is Ecopoetics?" in *Jacket2*.

21. Emerson, *EP*, 1042. On the "chthonic," see Haraway, *Staying with the Trouble*, 31, 53.

22. Emerson, *JMN*, 16:183–84.

23. Emerson, *JMN*, 4:199–200.

24. Emerson notes later: "I heard on the day when I graduated at Cambridge, in 1821, Sampson Reed, who on that day took his Master's degree, delivered his oration on Genius. It was poorly spoken, as A. Adams said, 'in a meeching way,' & the audience found it very dull & tiresome. . . . But I was much interested in it, &, at my request, my brother William, of Reed's class, borrowed afterwards the manuscript (of him for me), & I copied the whole of it, & kept it as a treasure. . . . In 1851, after Webster's Fugitive Slave Law, I met Reed in the Boston Atheneaum, & deplored to him this downfal [*sic*] of our great man, &c. He replied, that 'he thought it his best speech, & the greatest action of his life.' So there were my two greatest men both down in the Pit together." Emerson, *JMN*, 16:184. Emerson may be conflating the actual sequence of events here. In 1821, Emerson graduated from Harvard's undergraduate class as a young eighteen-year-old, just a few days after the commencement exercises for the Master's class where Reed had spoken. See R. Richardson, *Mind on Fire*, 16–17.

25. *EP*, 1291–92.

26. Reed, *SR*, 16.

27. Reed, *SR*, 13.

28. Reed, *SR*, 15.

29. Reed, *SR*, 14–15.

30. Walls, *Emerson's Life in Science*, 75.

31. This is Sylvia Shaw's insightful claim. Shaw, "Sampson Reed," 48, 68.

32. Emerson, *EP*, 676.

33. Emerson, *EP*, 1291–92.

34. Reed, *SR*, 24.

35. Reed, *SR*, 31.

36. Walls, *Emerson's Life in Science*, 98.

37. Reed, *SR*, 35.

38. Reed, *SR*, 33.

39. My emphasis; Reed, *SR*, 34.

40. See Cameron, *Emerson*, 1:289; and Shaw, "Sampson Reed," 86–87.

41. Given Reed's specificity here about the smallest blades of grass being let to speak for themselves, it is tempting to hazard a direct connection between Reed and Whitman. While no specific correlation between the two has, to my knowledge, ever been precisely located, both David Reynolds and Anders Hallengren have read Whitman's general immersion in Swedenborgian theology as centrally important to his mature poetics. See Reynolds, *Walt Whitman*, 94–100; and Hallengren, "Hermeneutic Key," 45–59.

42. Reed, *SR*, 41.

43. For example: "Swedenborg's universe is a gigantic crystal, all whose atoms and laminae lie in uninterrupted order, and with unbroken unity, but cold and still," and also, "his varied and solid knowledge makes his style lustrous with points and shooting spicula of thought, and resembling one of those winter mornings when the air sparkles with crystals." Emerson, *EP*, 682, 668. Eric Wilson pays careful attention to the nexus between

Swedenborg's crystallography and Transcendentalists such as Emerson and Thoreau in *Spiritual History of Ice;* see also Richardson, *Natural History of Pragmatism,* 75–76.

44. Abram, *Spell,* 52.

45. Jacob von Uexküll's *Umwelt* critically developed the earlier insights of Ernst Haeckel, who famously coined, as discussed in this book's introductory chapter, our modern understanding of "ecology." The *Umwelt,* however, is not a holistic and integrated system of relations "out there" in the natural world—as Haeckel's *Ökologie* ostensibly is—but rather the symbolic (and semiotic) perception of the total biological relations between a being and other entities around it. Uexküll emphatically introduces an element of feeling subjectivity as a key constitutive part of the *Umwelt,* even for something as small as a tick—"time, which frames all happening, seems to us to be in the only objectively stable thing in contrast to the colourful change of its contents, and now we see that *the subject sways the time of his own world.* Instead of saying, as heretofore, that without time, there can be no living subject, we shall now have to say that without a living subject, there can be no time." Uexküll, "A Stroll through the Worlds of Animals," 37–38. As Wendy Wheeler succinctly and useful puts it, "An *Umwelt,* in other words, is a space of semiosis." Wheeler, *Whole Creature,* 103.

46. Emerson, *EL,* 130.

47. See Bosco, "His Lectures," 27.

48. Emerson, *Later Lectures,* 1:166.

49. Matthews, *For Love of* Matter, 2, 101–12.

50. Osborne, "Bad Seeds," 58.

51. Bennett, *Vibrant Matter,* 45–47.

52. My emphasis; Emerson, *Later Lectures,* 1:158.

53. Skrbina, *Panpsychism,* 224.

54. Emerson was one of Humboldt's "earliest and most steadfast American readers," according to Walls, even though Emerson did not engage with Humboldt's critiques of Idealism that were ultimately antithetical to Emerson's own position. Walls, *Passage to the Cosmos,* 251–55.

55. Emerson, *EP,* 1039.

56. Emerson, *EP,* 1040; Emerson's thinking of semiotics as akin to the activity of coral-reef building mirrors his earlier rhetoric in "The Poet" (1844): "Language is fossil poetry. As the limestone of the continent consists of the infinite masses of the shells of animalcules, so language is made up of images, or tropes, which now, in their secondary use, have long ceased to remind us of their poetic origin." Emerson, *EP,* 457.

57. "In loyalty both to the rules of interpreting scripture and to the mother tongue, which is part of our condition, the writer's words must on the first level make literal or historical sense, present the brutest of fact. It is that condition from which, if we are to hear significantly, 'we must be born again.' A son of man is born of woman; but rebirth, according to our Bible, is the business of the father. So *Walden*'s puns and paradoxes, its fracturing of idiom and twisting of quotation, its drones of fact and flights of impersonation—all are to keep faith at once with the mother and the father, to unite them, and to have the word born in us." Cavell, *Senses,* 16.

58. Osborne, "Bad Seeds," iii. Osborne also usefully notes how Reed's work anticipates the object-oriented ontology of Timothy Morton and Graham Harman, 138.

59. Reed, *SR*, 13.

60. L. Wilkinson, *Dream*, 127. Wilkinson argues that Oegger's texts were instrumental in shifting Swedenborgian correspondence theory into an "urban aesthetic" that later culminates in the "*correspondance*" of Baudelaire. See the discussion in the chapter 2, 56–57.

61. Emerson, *JMN*, 5:60–70.

62. In the "Language" chapter of *Nature*, Emerson repeats Oegger's idea that "primitive man" had a closer relationship with nature, and that words were once one with things: "Because of this radical correspondence between visible things and human thoughts, savages, who have only what is necessary, converse in figures. As we go back in history, language becomes more picturesque, until its infancy, when all is poetry; or all spiritual facts are represented by natural symbols." *EP*, 22.

63. Emerson, *JMN*, 5:77–78.

64. Just to be clear for the sake of the variegated history of translations of Swedenborg into English: *A Theosophic Treatise on the Nature of Influx* was a title that appeared only in rare late eighteenth- and early nineteenth-century English translations—later translations of *De Commercio Animae et Corporis*, from Emerson's day up to the present, have opted for either *Intercourse between the Soul and Body* or *Interaction between Soul and Body*—losing, in a sense, the two terms of "theosophy" and "influx" that carried so much cachet within an esoteric vitalism that was essential for germinating transatlantic Romanticism.

65. Swedenborg, *Theosophic Treatise*, no. 11.

66. As noted by Cameron, *Emerson*, 1:230, 236, 239, and 250. See also Emerson, *JMN*, 4:288, 342. Clothing as a metaphor for spirit, of course, is one of the guiding symbols in Thomas Carlyle's *Sartor Resartus* (1831)—Emerson makes occasional reference to "affections clothed" (and Sampson Reed's terminology, wishing Carlyle would "make his words one with things") in his letters to Carlyle in this period. Emerson, *CEC*, 99.

67. This is Hallengren's observation in *Code*, 130.

68. Emerson, *EP*, 196.

69. Swedenborg, *HH*, no. 26. Swedenborg's celestial angels are occasionally described as living nomadic, pastoral existences in heaven, living in tents and abstaining from eating meat. These details went onto be evoked in nineteenth-century communitarian countercultures, such as Fourierism and Associationism. Swedenborg's presence in early vegetarian movements is noted by Block, *New Church*, 68, 79. Such contexts are richly satirized in Nathaniel Hawthorne's short story, "The Hall of Fantasy" (1843), where the companion of the narrator who has spoken reverently of Swedenborg's statue in the imagination's veritable Hall of Fame, interrupts (and ends) the narrative with his need for a vegetarian dinner of vegetables and fruit. Hawthorne, *Tales*, 744–45. For Hawthorne's relation to nineteenth-century Swedenborgianism, see Zuber, "Spiritualized Science," 175–216.

70. Hallengren, *Code*, 130.

71. As quoted by Emerson, *JMN*, 1:195–196.

72. Hallengren, *Code*, 131.

73. This is David Abram's reading of Merleau-Ponty. Abram writes of the "Flesh" as a perceptual field for environmental ethics: situating "the reciprocal presence of the sentient in the sensible and of the sensible in the sentient, a mystery of which we have always, at least tacitly, been aware, since we have never been able to affirm one of these phenomena, the perceivable world or the perceiving self, without implicitly affirming the existence of the other." Abram, *Spell*, 66.

74. Richardson artfully traces "the fact of feeling" in *Natural History of Pragmatism*, reading Emerson's writing as an "oscillation between the poles of natural fact and religious feeling." *Natural History of Pragmatism*, 78.

75. Emerson, *EP*, 478.

76. Arsić, *On Leaving*, 14. Cameron's well-known reading of Emerson and the impersonal is in *Impersonality*, 79–107, as well as "Way of Life by Abandonment," 3–40. Cavell's thinking on Emerson in this regard is primarily in *Conditions Handsome and Unhandsome*, 33–63.

77. Richardson, *Natural History of Pragmatism*, 94.

78. My emphasis; Emerson, *EP*, 256.

79. Emerson, *CEC*, 32–33.

80. Emerson, *JMN*, 5:481; 7:31; 9:300; 10:150, 357.

81. Cf. "As for King Swedenborg I object to his cardinal position that evils should be shunned as sins. I hate preaching. I shun evils as evils. . . . As for this 'shunning evils as sins,' I prefer the ethics of the Vishnu. Too much form, O Swedenborg! too many steps, too much dogma, too much government. I see not that the soul can have need of elegance, said Confucius." Emerson, *JMN*, 9:313–14.

82. Harding, "Swedenborgian Spirit," 70.

83. Ibid., 66.

84. Wilkinson was at the center of a transatlantic nexus between Concord and London—he served as homeopathic physician for the Nathaniel Hawthorne family during their European sojourn ("Lizzie" Siddall, Dante Gabriel Rossetti's lover, and then wife, was also a patient); for literary historiography, Wilkinson remains notable for bringing out the first letterpress edition of William Blake's *Songs of Innocence and Experience*, in 1839. An extensive collection of letters by Wilkinson to Emerson, Henry James Sr., and others, now held at the London Swedenborg Society, evince how critically Wilkinson served to mediate American Transcendentalist understandings of Swedenborg. See, in particular, James John Garth Wilkinson, boxes K/123 (Letters to Henry James Sr.), E 1–8 (Letters to Ralph Waldo Emerson), 124a (Letters to Emma Anne Marsh), SSL. Wilkinson's significance to Transcendentalism is substantially recovered in recent books by Haller, *Distant Voices;* and Peet, *Medicine, Mysticism, and Mythology.*

85. Emerson, *EP*, 670.

86. Emerson, *Letters*, 3:342.

87. An even earlier date is possible. While the extant letters between Emerson and Wilkinson start in 1846—in a marvelous cache of letters in the archives of the London

Swedenborg Society—other letters from Wilkinson to his wife make it clear that starting in 1838, Wilkinson maintained a semi-regular correspondence with Thomas Carlyle. In 1838, Wilkinson had sent Carlyle a packet of books containing several Swedenborg titles, as well as Wilkinson's reprint of William Blake's *Songs of Innocence and Experience*. Carlyle responded in kind to Wilkinson, and noted: "Hitherto I have known nearly nothing of Swedenborg; or indeed I might say less than nothing, having been wont to picture him as an amiable but inane visionary with affections quite out of proportion to his insights. . . . But I have been rebuked already: a little Book by one Sampson Reed of Boston in New England which some friend sent hither, taught me that a Swedenborgian might have thoughts of the calmest sort on the deepest things." This "some friend" who had sent on Reed's "little Book hither" was, of course, Emerson, who had sent Carlyle a copy of *Growth of the Mind* in 1834. Carlyle and Carlyle, *Collected Letters* (August 2, 1838) 10:141–42.

88. Hotson, "Emerson's Biographical Sources for 'Swedenborg.'" Hotson's voluminous PhD dissertation written at Harvard in 1929 is still, perhaps, the most exhaustive historical account of Emerson's reading of Swedenborg, even as it overstates the case for Swedenborg as a singular sort of influence. Hotson, "Emerson and Swedenborg."

89. See Odhner, *Wilkinson*, 19.

90. Kroeber, *Ecological Literary Criticism*, 58.

91. J. J. G. Wilkinson, *Swedenborg*, 261–62.

92. Emerson, *EP*, 670.

93. Harding, "Swedenborgian Spirit," 71.

94. J. J. G. Wilkinson, "Correspondence," 138.

95. Ibid., 135–36.

96. Emerson, *Letters*, 3:414.

97. After receiving several letters and packets with materials from Wilkinson in 1838, Carlyle remarked in his journal: "Strange correspondence from Swedenborgians here, especially from one Wilkinson—a curious, shy proud deep hearted simple kind of man seemingly. Wishes me to write on Swedenborg. Begin to actually form some outline of a notion of Swedenborg. The eighteenth century: mechanicals and spirituals: the spirit of man struggling beneath loads of metallurgy, practicals, cause and effect, and etcetera. Swedenborg, Voltaire." See source note, TC to James John Garth Wilkinson, Carlyle and Carlyle, *Collected Letters* (September 28, 1839) 11:189–90.

98. J. J. G. Wilkinson, "Introductory Remarks," 6.

99. As the transatlantic correspondence of Henry James Sr. and Wilkinson makes abundantly clear, taking Swedenborg as a philosopher of ideas (and not ostensibly as theologian or exegete) prompts both Wilkinson and James (especially in the latter's *Substance and Shadow*) to shape unusually sharp Kantian critiques that subsequently left a marked impact on Pragmatism. K/123, SSL.

100. Emerson, *EP*, 771.

101. Reed, *SR*, 24, 29.

102. Thoreau, *Journal*, 4:46.

103. Gura, *Wisdom*, 137.

104. Muir, *Nature Writings*, 296.

4. Heralds of a New Gospel

1. M. P. Cohen, *History of the Sierra Club.*

2. Fox, *American Conservation Movement*, 368.

3. Emerson, *LRE*, 5:201–4.

4. Muir and Carr, *Kindred*, 198.

5. Starting with the unpublished memoir by Alfred Worcester (Joseph Worcester's nephew and closest relative), a false claim has been consistently repeated that as Worcester worked away at his "churning and chores" for Hutchings in 1864, he came to make Muir's acquaintance. This is impossible, as Muir remained in remote corners of northeastern Canada that summer, and did not get to Yosemite Valley himself until 1869 (the formative year that later became Muir's *My First Summer in the Sierra*). Alfred Worcester, "Rev. Joseph Worcester," SFSC, 3.

6. Freudenheim, *Building with Nature*, 1–3.

7. Limbaugh notes that Keith tried "without much success to encourage Muir to attend." Limbaugh, "California's Kindred Spirits," 79.

8. See, for example, Michael Cohen's *Pathless Way;* Donald Worster, in his definitive biography of Muir, sees Muir as a kind of secularizing agent, "cutting away at the roots" of old Christian theology (*Passion for Nature*, 416). As John Gatta and Evan Berry have shown, such approaches neglect the deep ways a Christian rhetoric continued to undergird Muir's beliefs. Gatta, *Making Nature Sacred*, 151–54; Berry, *Devoted to Nature*, 79–80. Muir's pantheistic tendencies, however, moved him into very heterodox positions (his belief in the souls of animals, for example), and Stephen Fox seems right in placing Muir within the "middle kingdom of psychic phenomena," such as spiritualism, that mediated between Christian and orientalist strands of nineteenth-century belief. Fox, *American Conservation Movement*, 368.

9. D. Williams, "John Muir," 83.

10. Muir, *Nature Writings*, 91.

11. Cronon, ed., *Uncommon Ground;* see, more recently, Callicott and Nelson, eds., *Wilderness Debate* and *Wilderness Debate Rages On*.

12. Budd, "American Background," 27–31.

13. Qtd. in Limbaugh, *John Muir's 'Stickeen,'* 27.

14. Stoll, "Milton in Yosemite," 237–74.

15. Mitchell, "Imperial Landscape," 5–33; see also A. Miller, "Fate of Wilderness," 94, and Brownlee, *Manifest Destiny*, passim.

16. P. Miller, *Nature's Nation*, passim; see also P. Miller, *Errand into the Wilderness*, chap. 9, "Nature and the National Ego," 204–16.

17. Harrison, "Art of William Keith," 29; see also Mary McHenry to Father Cornelius, Oct 1, 1938. C-B 595, Carton 5, BLUC.

18. Limbaugh, "California's Kindred Spirits," 77. Limbaugh sees Keith and Muir as the veritable inheritors of the American "poet-painter" trope made so famous by Asher Durand's painting *Kindred Spirits* (1849), which depicts the Hudson River School painter Thomas Cole and the poet William Cullen Bryant conversing together as they gaze on a wilderness prospect.

19. Limbaugh, "California's Kindred Spirits," 68.

20. L. M. Wolfe, *Son of Wilderness,* 190.

21. Pauley, "The Importance of John Muir's First Public Lecture."

22. The phrasing is Cole's; see Wallach, "Wadsworth's Tower" and "Some Further Thoughts" for a discussion vis-à-vis Emerson's famous "transparent eyeball" of Nature and a Foucauldian concern with nineteenth-century power and surveillance.

23. My emphasis; Muir, *Nature Writings,* 245.

24. Ibid., 211.

25. Ibid., 577.

26. On the thingness of natural phenomena (in contradistinction to their functioning as objects—a difference much belabored by Martin Heidegger), see Brown, *Other Things,* 17–48.

27. Derrida, *Writing and Difference.*

28. On Humboldt's importance for the aesthetics of Church and Bierstadt, see Walls, *Passage to the Cosmos,* 135, 320–23.

29. Neuhaus, *William Keith,* 20–22.

30. Only Keith's letters to Worcester survive, as Worcester was largely successful in destroying all of his private diaries, papers, and letters before he died. "Letters of William Keith to Rev. Joseph Worcester," Keith-McHenry-Pond Papers, box 1, BLUC.

31. According to notes in the papers of Keith's wife, these books in Keith's library included: Benjamin Worcester, *The Life and Mission of Emanuel Swedenborg* (Boston, 1883); James John Garth Wilkinson, *Swedenborg: A Biographical Sketch* (London, 1886); Theophilus Parsons, *The Religion and Philosophy of Swedenborg* (Boston, 1876); Swedenborg, *The Four Leading Doctrines of the New Church* (New York, 1857); a twelve-volume set of the *Arcana Caelestia;* Benjamin Worcester, *Swedenborg: Harbinger of the New Age of the Christian Church* (Philadelphia, 1910); Richard McCully, *Swedenborg Studies* (London, 1875); two copies of Swedenborg's *Spiritual Diary* (the nineteenth-century translated title of what is now called *Spiritual Experiences*); *Divine Attributes from the Apocalypse Revealed of Emanuel Swedenborg* (Philadelphia, 1866); Swedenborg, *Angelic Wisdom Concerning the Divine Love and Wisdom* (New York, 1890); Swedenborg, *The Doctrine of Faith and Charity from the Arcana Caelestia* (Boston, 1868); Swedenborg, *Angelic Wisdom Concerning the Divine Providence* (London, 1886); Swedenborg, *Lessons in Correspondences* (New York, 1892); and also Tafel's massive *Documents Concerning the Life and Character of Emanuel Swedenborg* (Inez Henderson Pond Collection, Series III, box 1, folder 18, UoP).

32. Harrison, "Art of William Keith," 63.

33. Ibid., 66.

34. Though scholarship on Keith has not substantively commented on his painting's departure from topographical specificity, it has been remarked upon as a striking characteristic of Inness's later work. See DeLue, *George Inness*, 243n33.

35. My emphasis; Keith, "Lecture by W. Keith, 1894 or 5," UoP.

36. My emphasis; qtd. in Harrison, "Art of William Keith," 60.

37. Even in the 1930s, critics like Neuhaus were observing how "under the influence of Worcester, Keith came to realize that what determines the lasting appeal of a painting is not so much what you paint as how you explore and exploit your subject-matter in terms of the purely abstract elements of design and color." Neuhaus, *William Keith*, 24.

38. Kandinsky's *Über das Geistige in der Kunst* (Concerning the spiritual in art), published the same year Keith died (1911), cemented synesthesia as a key part of the Modernist turn toward abstraction. For an examination of how Swedenborgian correspondence theory, mediated through French Symbolism, came to impinge on Kandinsky's ideas, see Walker, "Language of Form and Color." Keith's words here are quoted in Cornelius, "William Keith, Rembrandt of Landscape," 70–71.

39. Keith, "Lecture by W. Keith, 1894 or 5," UoP.

40. Qtd. in Limbaugh, "California's Kindred Spirits," 69.

41. Keith, "Lecture by W. Keith, 1894 or 5," UoP.

42. Worster, *Passion for Nature*, 374.

43. Muir's copy of J. J. G. Wilkinson, *Emanuel Swedenborg*, 278, John Muir Archive, UoP.

44. Toward the end of his life, Muir collected many articles on spiritualism, and dictated a section for his autobiography entitled "Mysterious Things" where he recounts three distinct incidents of his inexplicable clairvoyance or premonition. See Fox, *American Conservation Movement*, 369.

45. Muir, *Nature Writings*, 766.

46. See the foregoing remarks on pp. 1–2.

47. Benjamin, *Illuminations*, 261.

48. Milton, *Paradise Lost*, 9.

49. Solnit, *Savage Dreams*, 222.

50. Righter, *Battle*, especially 191–215.

51. Keith seems to have created between thirty-eight and forty landscape sketches of Hetch Hetchy out of these two respective trips. Brewster, *Comprehensive Keith*, 198.

52. A. Miller, "Fate of Wilderness," 101–2.

53. On Bierstadt's painterly distortions and exaggerations, see Green, *Carlton Watkins*, 365–68. Most (in)famously, Bierstadt went so far as to (falsely) christen a Sierra Nevada mountain in a painting "Mount Corcoran" to entice the wealthy patron William Wilson Corcoran to buy the work. Lisa Strong, "Mount Corcoran," 140.

54. Keith, "Lecture by W. Keith, 1894 or 5," UoP.

55. Swedenborg, *AC*, nos. 794, 847.1

56. See Brechin, *Imperial San Francisco*, 101–17, for a compelling reading of the dam advocacy as fueled by greed and private wealth, rather than public interest.

57. *Congressional Record,* 5467. Quite ironically, during the debate on the floor of the Senate, the prominent figuration of water in Keith's painting(s) led to a prodam advocate's (Nevada senator Key Pittman) pointing out how the presence of water in the valley would seem only to enhance its scenic potential, as Keith's paintings apparently evinced.

58. Muir, *Nature Writings,* 817.

59. On the central role of a kind of realist photography for the Sierra Club and other early conservation efforts, see Dunaway, *Natural Visions,* 3–30; and Solnit, "Every Corner," 213–36.

60. See Green, *Carlton Watkins.* Green also makes a highly persuasive argument for viewing Watkins as continuing a distinctly Emersonian aesthetic, 57–70.

61. Green, *Carlton Watkins,* 281.

62. Muir, "Looking Up Hetch-Hetchy [*sic*] Valley from Sunrise Point," in *Yosemite,* 204; Muir, "Hetch-Hetchy [*sic*] Valley," in *Yosemite,* 244.

63. For an excellent gloss on these tensions, see N. Howe, *Landscapes,* especially 149–52; as well as Burton, *Worship and Wilderness.*

64. See Righter's concluding discussion of the "Restore Hetch Hetchy" movement, *Battle,* 216–41.

65. Freudenheim, *Building with Nature,* xi–xii. The church definitively "heralds the arrival of Arts and Crafts tenets in northern California," according to Bosley, "A. C. Schweinfurth," 12.

66. Tobisch, "Garden Church." Freudenheim notes how "in Japan, as at the Swedenborgian church, sacred shrines are enclosed by high walls or fences and are entered by passing through a gate and then through a garden. The church garden's careful composition of blossoming plum and apple trees suggest the artful landscaping of Japanese gardens." *Building with Nature,* 47. I am indebted to my graduate student Austin Salzwedel for pointing out to me how this kind of design was also quite typical of Japanese-style tea houses, and that Worcester's self-designed home in the Piedmont hills (then unincorporated Oakland) was very near to the Japanese tea house that was constructed during the redevelopment of Piedmont Park in the 1890s. See Swift, *Bubbling Springs.*

67. Albanese, *Nature Religion in America* and *Reconsidering Nature Religion;* and B. Taylor, *Dark Green Religion,* particularly 48–70.

68. Berry, *Devoted to Nature,* 69–70.

69. Ibid., 66, 138.

70. "The slow-going, unthrifty farmers, also, are beginning to realize that when timber is stripped from the mountains the irrigating streams dry up in summer, and are destructive in winter; that soil, scenery, and everything slips off with the trees: so of course they are coming into the ranks of tree-friends." Muir, *Nature Writings,* 709.

71. Swedenborg, *HH,* no. 111.

72. Swedenborg, *AC,* no. 2722.

73. See the discussion in chapter 2, 69–70.

74. Bosley, "A. C. Schweinfurth"; Freudenheim, *Building with Nature,* passim.

75. Freudenheim, *Building with Nature*, 48.

76. Typescript of Ednah Silver, "Sketches of the New Church," in Worcester, Joseph, Box 72, folder 9, CSS. Silver notes how parishioners brought "Dryad treasures from weekend wood wanderings to the altars" of the church.

77. Adrienne Copa, "The Garden Church of San Francisco: A History" (n.d.), SFSC.

78. In his unfinished memoirs, published posthumously by William Badè, Muir wrote: "I boarded with Mr. Hutchings' family, but occupied a cabin that I built for myself near the Hutchings' winter home. This cabin, I think, was the handsomest building in the Valley, and the most useful and convenient for a mountaineer. From the Yosemite Creek, near where it first gathers its beaten waters at the foot of the fall, I dug a small ditch and brought a stream into the cabin, entering at one end and flowing out the other with just current enough to allow it to sing and warble in low, sweet tones, delightful at night while I lay in bed. The floor was made of rough slabs, nicely joined and embedded in the ground. In the spring the common pteris ferns pushed up between the joints of the slabs, two of which, growing slender like climbing ferns on account of the subdued light, I trained on threads up the sides and over my window in front of my writing desk in an ornamental arch. Dainty little tree frogs occasionally climbed the ferns and made fine music in the night, and common frogs came in with the stream and helped to sing with the Hylas and the warbling, tinkling water. My bed was suspended from the rafters and lined with libocedrus plumes, altogether forming a delightful home in the glorious Valley at a cost of only three or four dollars, and I was loath to leave it." Badè, ed., *Life*, 1:208–9.

79. Freudenheim, *Building with Nature*, 17.

80. See, for example, the stereoscopic image "Room in the Old Hutchings House," Underwood & Underwood., accessed February 20, 2018, http://www.loc.gov/pictures/resource/cph.3c00937/.

81. "The Myth of the Tree You Can Drive Through," National Park Service, accessed March 1, 2018, https://www.nps.gov/seki/faqtunnel.htm.

82. Muir, *Nature Writings*, 687, 813, 745.

83. Solnit, *Savage Dreams*, 307–8; see also Spence, *Dispossessing the Wilderness*, 101–13.

84. Solnit, *Savage Dreams*, 229.

85. Muir, *Nature Writings*, 726.

86. Weintraub, *To Life!*, 183–88; religious studies scholars Jeffrey Kosky and Mark Taylor further claim Goldsworthy's work as a critical site of reenchantment and post-secular spirituality; M. C. Taylor, *Refiguring the Spiritual*, 156–79; Kosky, *Arts of Wonder*, 131–66.

87. Benjamin's *Jetztzeit* appears in the dense, aphoristic meditations on history from 1940. The "here-and-now" of *Jetztzeit*, irrupting the "homogenous empty time" of modernity, does not occur naturally or teleologically on its own, but is instantiated by the aesthetic—causing "a tiger's leap into the past," a fusion of the horizons of past, present, and future. Benjamin, *Illuminations*, 261. For a broader reflection on aesthetics and time in Benjamin vis-à-vis public art at Ground Zero, see Zuber, "Flânerie."

88. See Erickson, "Theophanic Materiality," 203–20, for an insightful reading of Goldsworthy through a pantheism animated by the new materialisms of Karen Barad and Jane Bennett.

89. See J. Rose, "Ornaments," for a full catalogue of Swedenborg's typographic illustrations and engravings.

90. Fountains or birdbaths "suggest the flowing center of the divine primordial garden, reflecting the numinous, life-giving center in one's own innermost self, the linking with which is the goal of processes of religious seeking and initiation." Ronnberg and Martin, eds., *Book of Symbols*, 608.

91. Swedenborg, *TCR*, no. 77; *HH*, no. 122.

92. Hanegraaff, *New Age Religion*, 427–28.

93. London, *Jack London*, 1:361.

94. My emphasis; qtd. in Brewster, *Comprehensive Keith*, 50.

95. Freudenheim, *Building with Nature*, 102–3.

96. Cornelius, *Old Master*, 27.

97. Harrison, "The Art of William Keith," 35. Muir's writing from the same period takes note of the devastating consequences of mining, but is also, surprisingly, ambivalent; California went on to ban hydraulic mining in 1884, and Limbaugh posits that Muir didn't really develop an environmental conscience until later. Limbaugh, "John Muir and the Mining Industry," 62.

98. See pp. 42–44.

99. A "rights of nature" for trees could be said to commence with Christopher Stone's landmark text, *Should Trees Have Standing?* (first published in 1973), but such a modern framework is incipient in Muir as early as 1867, when Muir ponders "the rights of all the rest of creation"—see Nash, *Rights of Nature*, 6, 38–41.

100. Muir, *Nature Writings*, 720.

101. Wohlleben, *Das geheime*, especially chap. 2, "Die Sprache der Bäume" (The language of the trees), 14–20. As Haraway notes, reflecting on the circuits of communication between squids and bioluminescent bacteria, "material semiotics is exuberantly chemical; the roots of language across taxa, with all its understandings and misunderstandings, lie in such attachments." Haraway, *Staying with the Trouble*, 66.

102. See the prior discussion on Uexküll on p. 187n45.

103. Muir, *Nature Writings*, 472–73.

104. Ibid., 245.

105. Ibid., 554.

106. Ibid., 568.

107. Ibid., 551.

108. Ibid., 571.

109. By "chthonic," I mean to reference Haraway's work as a multispecies theorist, her insistent attention to "beings of the earth," and consequent proposal to redescribe the Anthropocene as "Chthulucene." Although Haraway rejects any kind of residue of transcendental "sky gods" religions with their inevitable dualisms, her Chthulucene is

weirdly resonant with a kind of postsecular tenor: "This Chthulucene is neither sacred nor secular; this earthly worlding is thoroughly terran, muddled, and mortal—and at stake now. . . . We are humus, not Homo, not anthropos; we are compost, not posthuman . . . unlike either the Anthropocene or Capitolocene, the Chthulucene is made up of ongoing multispecies stories and practices of becoming-with in times that remain at stake, in precarious times, in which the world is not finished and the sky has not fallen—yet." Haraway, *Staying with the Trouble*, 55.

110. Bron Taylor sees this moment in Muir as critically "presaging" Arne Naess's "ecological self," *Dark Green Religion*, 65–66.

111. My emphasis; Muir, *Nature Writings*, 826.

112. Bennett, *Vibrant Matter*, 61.

113. Luciano, "Sacred Theories," 717. Luciano only briefly cites Swedenborg's importance for the advent of American spiritualism; according to Leigh Eric Schmidt, Swedenborgian doctrines created an elaborate "spiritual sensorium" of perception that came to shape much American spiritualist discourse, beginning with seers like Andrew Jackson Davis. Schmidt, *Hearing Things*, 199–215.

114. Luciano, "Sacred Theories," 726–27.

115. Wurtz, "John Muir Disrupts a Séance," 3–4.

116. In a hand that seems to be Muir's, a number of passages pertaining to Swedenborg's clairvoyance and preternatural abilities are found in Muir's copy of J. J. G. Wilkinson's Swedenborg biography (London: Speirs, 1886), 90–91, 103, 122–24, 126, 128. John Muir Archive, UoP.

117. Muir, *Travels in Alaska*, 151–52.

118. Swedenborg, *DLW*, no. 157; Swedenborg, *DP*, no. 233, sec. 10.

119. A sympathetic (if brief) reading of Muir's relationship to Native Americans is offered by Fleck, *Henry Thoreau and John Muir*, 29–80. Solnit and others have been more critical of Muir's characteristic nineteenth-century racism; *Savage Dreams*, 284–87. A nuanced perspective is offered by Daniel Lee Henry, who examines Muir's Alaskan trip from the perspective of the Tlingit people, reading Muir's arrival as a harbinger of modernity and all its complications for the indigenous tribes he encountered. Henry, *Across the Shaman's River*.

120. Muir, *Nature Writings*, 372–73.

121. Ibid., 571.

122. For a discussion of some of these struggles and how Native American literature and art has negotiated legal regimes, see Zuber, "Reservation Paints Back."

5. Homes for Herons

1. Jewett and Inness both published, for example, in *Harper's New Monthly Magazine*—Inness with his article "A Painter on Painting" (November 13, 1867), and again with several poems in 1882, and Jewett with six stories between 1892 and 1902. Jewett was also probably aware of Inness's article on color that had appeared in the Swedenborgian periodical the *New Jerusalem Messenger* (also in 1867), as her mentor

Theophilus Parsons refers her to a text by Dr. Bayley—"The Ribband of Blue"—that precipitated Inness's longer article in the *Messenger.*

2. Graham and Buchheister, *Audubon Ark,* 3–13.

3. Orr, *Saving American Birds,* 29, 52; see also Merchant, *Spare the Birds!,* 27.

4. Philippon, *Conserving Words,* 1–30.

5. DeLue, *George Inness,* 2, 21; Promey, "Ribband," 44–48.

6. In answer to a question from an interviewer about what happens when he grew tired of painting, Inness replied: "Then I take to theology. That is the only thing except art which interests me. In my theory, in fact, they are closely connected. That is, you may say it is theology, but it has resolved itself into a scientific form, and that is the development which has become so very interesting to me. I have written piles upon piles of manuscripts upon it, and my method is to take these piles and rewrite them in very condensed form. Gradually this grows and is 'boiled down,' and all the first essays destroyed." Inness, *Writings,* 93.

7. Emerson, *EP,* 475.

8. Cikovsky, *George Inness,* 107.

9. Marx, *Machine in the Garden;* and Fletcher, *New Theory.*

10. Marx, *Machine in the Garden,* 23, 71; A. Miller, "Fate of Wilderness," 93, 100; see also Novak, *Nature and Culture,* 149–52.

11. Schuyler, *Sanctified Landscape,* 1–27.

12. Marx, *Machine in the Garden,* 220.

13. Cikovsky reads *The Lackawanna Valley* as not being critical in any way of industrialization and progress, "but rather condones, even glorifies, the situation [Inness] represents." Cikovsky, "George Inness and the Hudson River School," 52; Cikovsky later updated and expanded on his claims for this painting in Cikosvky, "'The Ravages of the Ax,'" 611–26. Novak offers a counter-reading, however, finding the painting "one of the most puzzling pictures in American art," with deeper ambivalence than Cikovsky and Marx allow for. Novak, *Nature and Culture,* 149–51.

14. Whitman later praises the locomotive in the 1876 version of *Leaves of Grass,* in a poem called "To a Locomotive in Winter," where the train is "Type of the modern—emblem of motion and power—pulse of the continent." Whitman, *Poetry and Prose,* 583.

15. Cikovsky, *George Inness,* 32; it seems likely this Mexico excursion occurred during his trip to California in 1891, the same period when he spent substantial time with William Keith in San Francisco (and Monterey), sharing a painting studio.

16. Marx, *Machine in the Garden,* 221.

17. Swedenborg, *DLW,* no. 61.

18. See the prior discussion in chapter 2, 69–70.

19. Inness, *Writings,* 68.

20. In her fascinating and careful attention to Inness's later use of scratches and scrub marks on the surface of his paintings, Rachel DeLue argues that these graphemes loosely functioned as a kind of deliberate evocation of writing—"Inness's scratches and scrapings push for the expansion, transformation, and collapse of categories, or

systems of denotation, usually understood as distinct; language is asked to become painting, and painting a form of language." DeLue, *George Inness*, 230.

21. The grant amount was $10,000, an extraordinary sum in 1867 (the approximate equivalent of over $150,000 in today's value). See DeLue, *George Inness*, 147.

22. Bell, *George Inness*, 82.

23. The structural collapse not only destroyed artworks in the gallery, but also killed four people and injured many more, and was widely reported in newspapers of the day. See, for example, "The Madison Avenue Ruin," *New York Times*, April 23, 1880.

24. Carol Vogel, "New Holdings in the Modern," *New York Times*, November 12, 2004.

25. On Blake and Swedenborg, see Paley, "'A New Heaven,'" as well as Paley's introductory commentary to Blake, *Jerusalem: The Illuminated Books*, 9–16.

26. Qtd. in DeLue, *George Inness*, 159.

27. Promey, "Ribband," 54.

28. My emphasis; Inness, *Writings*, 113.

29. Novak, *Nature and Culture*, 149—see also note 13 above.

30. Cole, "Essay on American Scenery," 3, 12.

31. Stoll, *Inherit the Holy Mountain*, 87–88. David Lowenthal also notes that next to Darwin's *Origin*, Marsh's *Man and Nature* was "the most influential text of its time to link culture with nature, science with society, landscape with history." Lowenthal and Cronon, *George Perkins Marsh*, xv.

32. Marsh, *Man and Nature*, 234.

33. Stoll, *Inherit the Holy Mountain*, 89; for more on the central importance of Humboldtian science to Marsh, see Walls, *Passage to the Cosmos*, 295–301.

34. Qtd. in DeLue, *George Inness*, 159.

35. As recorded by Burnham's assistant, Edward Bennett: "We went to see Dr. Worcester, Swedenborgian minister of San Francisco, a life-long friend. Mr. Burnham always acknowledged his influence. 'He has always kept a hand on my shoulder' was the way he put it. The laws of spiritual correspondence were often in his mind and one evening at the bungalow on Twin Peaks he interested himself in tracing the correspondence of spiritual powers and the municipal powers as indicated in the physical lay-out of the centre of the city. I had proposed a system of concentric rings or circuits surrounding the ideal Centre of Government." Qtd. in Moore, *Daniel H. Burnham*, 2:170.

36. W. Wilson, *City Beautiful*, 57–59.

37. Fisher, "Iconology," 449.

38. See the prior discussion of the *Maximus Homo* in chapter 2, pp. 64–68. This point is made by Fisher, "Iconology," 451; and Schaffer, "Burnham's Plan of Chicago."

39. Schaffer, *Daniel H. Burnham*, 19.

40. Inness, *Writings*, 129.

41. DeLue, *George Inness*, 190.

42. Inness, *Writings*, 142.

43. Swedenborg, *HH*, nos. 191–199; and *DLW*, no. 69. The nineteenth-century esotericist Ethan Allan Hitchcock was one of the first to argue for substantial parallels between Swedenborg's immanental cosmology and Spinoza's pantheism. Hitchcock, *Swedenborg*, 263–331.

44. Block, *New Church*, 102. Joseph's father, Thomas, was the Harvard student who had heard rumors that copies of Swedenborg's *Arcana Coelestia* (donated to the college by William Hill at the end of the eighteenth century) lay somewhere on campus, and eventually (re)discovered the volumes covered with dust and trash in the "museum room" of the Harvard library, alongside stuffed crocodiles and other exotic specimens.

45. For a partial list of Jewett's extant library, see Harvard College Library, "List."

46. Sarah Orne Jewett to Theophilus Parsons, November 14, 1872. Transcription by Fraser Cocks. Sarah Orne Jewett Text Project, Colby College, accessed December 18, 2013, http://www.public.coe.edu/~theller/soj/let/parsons.html.

47. Jewett to Parsons, September 14, 1873, accessed December 18, 2013, http://www.public.coe.edu/~theller/soj/let/parsons.html.

48. See Donovan, "Jewett and Swedenborg," 732; Matthiessen also notes this in his *Sarah Orne Jewett*, 52.

49. Donovan, "Jewett and Swedenborg," 732.

50. S. Mayer, *Naturethik und Neuengland-Regionalliteratur*, 151–200.

51. Murphy, "Grounding Anotherness," 417–34.

52. Jewett, *Novels and Stories*, 55.

53. Donovan, "Jewett and Swedenborg," 736–37.

54. Jewett to Parsons, September 14, 1873, accessed December 18, 2013, http://www.public.coe.edu/~theller/soj/let/parsons.html.

55. Jewett, *Novels and Stories*, 16.

56. Swedenborg, *HH*, nos. 291–302, 599.

57. Jewett to Parsons, August 24, 1876, accessed December 18, 2013, http://www.public.coe.edu/~theller/soj/let/parsons.html.

58. Parsons, *Outlines*, 71.

59. Jewett to Parsons, October 25, 1874, accessed December 18, 2013, http://www.public.coe.edu/~theller/soj/let/parsons.html.

60. Emerson, *EP*, 10.

61. Jewett, *Novels and Stories*, 138.

62. Jewett to Parsons, June 8, 1877, accessed December 18, 2013, http://www.public.coe.edu/~theller/soj/let/parsons.html.

63. Fetterley and Pryse, *Writing out of Place*, 7, 29.

64. Jewett, *Novels and Stories*, 21.

65. Ibid., 140.

66. Emerson, *EP*, 266.

67. Jewett, *Novels and Stories*, 103–4.

68. "Kate laughingly proposed one evening, as we sat talking by the fire and were particularly contented, that we should copy the Ladies of Llangollen, and remove ourselves from society and its distractions." Jewett, *Novels and Stories*, 135. For a reading

of Lady Butler and Sarah Ponsonby's transgressive coupling via Judith Butler's work on queer melancholia, see Brideoake, "Extraordinary Female Affection."

69. I mean *communitas* here in the sense of creating a sacred sense of community, as deployed by Victor Turner and others within the field of pilgrimage studies. Turner, *Dramas, Fields, and Metaphors*, 273.

70. On Blake's use of Swedenborg in *The Marriage of Heaven and Hell*, see Viscomi, "In the Caves of Heaven and Hell," 27–60.

71. The parsing of biological sex from the cultural performance of gender is the baseline of Judith Butler's *Gender Trouble;* see especially 189–93 for her now-classic argument about how gender "ought not to be construed as a stable identity or locus of agency from which various acts follow; rather, [it] is an identity tenuously constituted in time, instituted in an exterior space through a *stylized repetition of acts.*" Swedenborg's gender binaries, conversely, seem at first to be construed as monolithic opposites—that men constitute the capacity for the reception of truth, and women embody love of (that) truth, which produces wisdom; and yet, it gets more complex (or at least confusing) when Swedenborg later states how "in the male the inmost is love and its clothing is wisdom, or what is the same, he is love veiled over with wisdom; and that in the female the inmost is that wisdom of the male, and its clothing is the love therefrom"—in other words, thus putting as the inner essence of each opposite sex what is ostensibly the other's "outer," what Butler would call the congealing of gender. Butler, *Gender Trouble*, 45; Swedenborg, *CL*, no. 32.

72. For the significance of Swedenborg for Balzac's *Seraphita* and *Louis Lambert*, see L. Wilkinson, *Dream*, 152–216.

73. Crowley, "'One [Hermaphroditic] Angel,'" 89–104.

74. Qtd. in Chevigny, *Woman and the Myth*, 112–13.

75. Emerson, *EP*, 680

76. Donovan, "Jewett and Swedenborg," 735.

77. Qtd. in Block, *New Church*, 157.

78. In addition to Josephine Donovan's important work, see also Sandilands, "Importance of Reading Queerly," 57–77.

79. Kermode claims that fiction excels at combining the diurnal "tick-tock" time of "uninteresting successiveness" (*chronos*) with the immanence of *kairos*-time, a transcendental significance that points toward a beginning or (apocalyptic) end, outside the frame of the diurnal: "[W]ithin this organization that which was conceived of as simply successive becomes charged with past and future: what was *chronos* becomes *kairos*. This is the time of the novelist, a transformation of mere successiveness which has been likened . . . to the experience of love, the erotic consciousness which makes divinely satisfactory sense out of the commonplace person." Kermode, *Sense of an Ending*, 46.

80. Jewett, *Novels and Stories*, 141.

81. Halberstam, *In a Queer Time and Place*, 1–2.

82. Matthiessen, *Sarah Orne Jewett*, 72.

83. Heise, *Imagining Extinction*, 23.

84. Jewett, *Novels and Stories*, 679.

85. On the "marriage plot" in theories of fiction and narrative form, from Ian Watt onward, see the recent excellent discussion by Seidman, *Marriage Plot*, 1–21.

86. For the complexities and paradoxes in Audubon's own nature aesthetics, see Zuber, "'Those Almost Uninhabited Shores.'"

87. Qtd. in Barrow, *Passion for Birds*, 141.

88. Merchant, *Spare the Birds!*, 24.

89. Ibid.

90. Jewett, *Novels and Stories*, 677–78.

91. See Quick, *Inness Catalogue*, 2:396.

92. "Slaughter in Florida," 139.

93. "Wholesale Destruction of Birds in Florida," 178–80.

94. Westfahl, *One Person Can Make a Difference*.

95. Merchant, *Spare the Birds!*, 27.

96. See Hartley, *Importance of Bird Life*, especially its appendix, "The Extermination of American Birds by Mankind," 302–5. Hartley's "The Last Parrakeet" was serialized in *St. Nicholas Magazine*, and then appeared as a small book in 1923. Hartley was not the only Inness descendant, it seems, to have been involved in ornithological conservation; his sister Rachel accompanied her brother with the leading zoologist William Beebe on bird expeditions in Guiana, where she painted specimens found in the jungles by her brother and the other men on the expedition. Raby, *American Tropics*, 76, 81.

97. For an arresting description of Eagleswood through the eyes of Thoreau, who found it a "queer place," see Walls, *Henry David Thoreau*, 392–93.

98. Thomas, *Alternative America*, 118.

99. For the prominent role of Swedenborgian theology in these different socialist-utopian projects, see Gladish, *Swedenborg, Fourier*.

100. The fullest consideration of Inness's engagement with George remains Leo Mazow in his doctoral dissertation, "George Inness: Problems in Antimodernism," and later related publications, such as "George Inness, Henry George."

101. Inness, *Writings*, 140.

102. My emphasis; ibid., 137.

103. Mazow, "George Inness, Henry George," 65–72.

104. Bell, *George Inness and the Visionary Landscape*, 124.

105. Qtd. in Quick, *Catalogue*, 2:396.

106. Heise, *Imagining Extinction*, 38.

107. C. Taylor, *Secular Age*, 26–28.

108. DeLue, *George Inness*, especially 21, 37, 83.

109. Bedell, *Moved to Tears*, 100–103.

110. Corn, "The Color of Mood," 229.

111. Inness, *Writings*, 136.

112. Heise, *Imagining Extinction*, 34.

Coda

1. Buhle and Van Sciver, *Johnny Appleseed*, "Introduction," n.p.

2. Powers, *The Overstory*, 161.

3. Ibid., 163.

4. Stengel, *Aufklärung bis zum Himmel*.

5. Blake, *Poetry and Prose*, 546.

6. Emerson, *JMN*, 13:335.

7. There are two recent strong exceptions to this: the Reverend Anna Woofenden's "Garden Church" that has drawn on the green roots of the Swedenborgian tradition to grow a vibrant spiritual community at the intersection of urban gardening and social justice concerns for the homeless in Los Angeles. (Woofenden is an ordained minister in the Swedenborgian Church of North America.) On the East Coast, Lincoln Smith has started a biodynamic "forest gardening" project on Swedenborgian church land in Maryland that aims to educate the public about sustainable consumption; Forested's spiral design and layout is specifically modeled on Swedenborg's vision of the heavenly garden in his *Conjugial Love* (email to author, December 11, 2012). For more on Woofenden (who studied for a year at the Center for Swedenborgian Studies in Berkeley) and the garden church, see Bass, *Grounded*, 46–48; Smith's Forested has been covered by the *National Geographic* in Pipkin, "How to Farm a Forest—and Feed a Neighborhood."

8. Pope Francis, *Laudato Si*.

9. Swedenborg has been cited as a central organizing figure for metaphysical traditions by Albanese, Bender, John Haller, Alfred Gabay, Arthur Versluis, and Gary Lachman, among others. See also J.-F. Mayer, "War Swedenborg der Herald des New Age?" 186–88.

10. On the formative role of Emersonian ideas for New Thought and the mind-cure movement, see Haller, *History of New Thought* and *Swedenborg, Mesmer*. Muir's radical legacy is accounted for by Pike, *For the Wild*, 43–44; and B. Taylor, "The Religion and Politics of Earth First!"

11. Haraway, *Staying with the Trouble*, 32.

12. Berry, *Devoted to Nature*, 188.

13. Tuan, *Topophilia;* the phrase "techniques of the observer" comes from Jonathan Crary's seminal exploration of modern ways of seeing, *Techniques of the Observer*.

14. This is somewhat addressed in chapter 2, see pp. 62–63. On Peirce's signal role for biosemiotics, see Wheeler, *Whole Creature*, 16–17, 32; and Maran, "Where Do Your Borders Lie?" 465–68.

15. Peirce, *Collected Papers of CSP*, vol. 8, par. 55–90.

16. Swedenborg, *DP*, no. 3, subsection 3.

17. Bennett, *Enchantment of Modern Life*.

18. Morton, *Dark Ecology*, 158.

19. Morton's weirdly weird is the "uncanny, unexpected fallout from the myth of progress"—the personalized, felt experience of realizing our part of being "a

geophysical force on a planetary scale" (a paraphrase of Chakrabarty). Morton, *Dark Ecology*, 7–9; see also Morton, *Ecological Thought*.

20. The car ignition example is Morton's; *Dark Ecology*, 8.

21. Meillassoux, *After Finitude*.

22. B. Taylor, *Dark Green Religion*.

23. Chakrabarty, "Climate of History," 220—see also *Whose Anthropocene?*, edited by Emmett and Lekan, particularly Chakrabarty's response, 103–13.

24. Ghosh, *Great Derangement*, 160–61.

25. Qtd. in Block, *New Church*, 117.

26. Lindsay, *Village Magazine*, 115.

27. Lindsay, *Johnny Appleseed*, 84–85. Much remains to be written on the specific ways that Lindsay's Modernist poetics channeled a strand of poetic Swedenborgianism that goes back to Emerson, via Whitman.

Bibliography

Abram, David. *The Spell of the Sensuous: Perception and Language in a More-Than-Human World.* New York: Pantheon Books, 1996.

Abrams, M. H. *Natural Supernaturalism: Tradition and Revolution in Romantic Literature.* New York: Norton, 1971.

Alaimo, Stacy. *Bodily Natures: Science, Environment, and the Material Self.* Bloomington: Indiana University Press, 2010.

———. *Exposed: Environmental Politics and Pleasures in Posthuman Times.* Minneapolis: University of Minnesota Press, 2016.

———. "Oceanic Origins, Plastic Activism, and New Materialism at Sea." In *Material Ecocriticism,* edited by Serenella Iovino and Serpil Opperman, 186–203. Bloomington: Indiana University Press, 2014.

Albanese, Catherine. *Nature Religion in America: From the Algonquin Indians to the New Age.* Chicago: University of Chicago Press, 1990.

———. *Reconsidering Nature Religion.* Harrisburg, Pa.: Trinity Press International, 2002.

———*A Republic of Mind and Spirit: A Cultural History of American Metaphysical Religion.* New Haven, Conn: Yale University Press, 2007.

Albert, Simon, Javier X. Leon, Alistair R. Grinham, John A. Church, Badin R. Gibbes, and Colin D. Woodroffe. "Interactions between Sea-level Rise and Wave Exposure on Reef Island Dynamics in the Solomon Islands." *Environmental Research Letters* 11, no. 5 (2016): 1–9.

Arsić, Branka. *On Leaving: A Reading in Emerson.* Cambridge, Mass.: Harvard University Press, 2010.

Badè, William Frederic, ed. *The Life and Letters of John Muir.* 2 vols. New York: Houghton Mifflin, 1924.

Barad, Karen Michelle. *Meeting the Universe Halfway: Quantum Physics and the Entanglement of Matter and Meaning.* Durham, N.C.: Duke University Press, 2006.

Barrow, Mark. *A Passion for Birds: American Ornithology after Audubon.* Princeton, N.J.: Princeton University Press, 2000.

Bass, Diana Butler. *Grounded: Finding God in the World—A Spiritual Revolution.* New York: HarperOne, 2017.

Bate, Jonathan. *The Song of the Earth.* Cambridge, Mass.: Harvard University Press, 2000.

Baudelaire, Charles. *Charles Baudelaire: Selected Poems*. Translated by Geoffrey Wagner. London: Falcon Press, 1947.

Bedell, Rebecca. *Moved to Tears: Rethinking the Art of the Sentimental in the United States*. Princeton: Princeton University Press, 2018.

Bell, Adrienne Baxter. *George Inness and the Visionary Landscape*. New York: National Academy of Design, 2003.

Bender, Courtney. *The New Metaphysicals: Spirituality and the American Religious Imagination*. Chicago: University of Chicago Press, 2010.

Benjamin, Walter. *The Arcades Project*. Translated by Howard Eiland and Kevin McLaughlin. Cambridge, Mass.: Belknap Press of Harvard University Press, 1999.

———. *Illuminations*. Translated by Harry Zohn. New York: Schocken Books, 2007.

Bennett, Jane. *The Enchantment of Modern Life: Attachments, Crossings, and Ethics*. Princeton, N.J.: Princeton University Press, 2001.

———. *Vibrant Matter: A Political Ecology of Things*. Durham, N.C.: Duke University Press, 2010.

Benz, Ernst. *Emanuel Swedenborg: Visionary Savant in the Age of Reason*. Translated by Nicholas Goodrick-Clarke. West Chester, Pa.: Swedenborg Foundation, 2002.

Bergquist, Lars. *Biblioteket i lusthuset: Tio uppsater om Swedenborg*. [The library in the summer house: Two essays on Swedenborg]. Stockholm: Natur och Kultur, 1996.

———. "Swedenborg and Linnaeus." *Arcana* 3, no. 2 (2004): 23–39.

———. *Swedenborg's Secret: The Meaning and Significance of the Word of God, the Life of the Angels, and Service to God*. London: Swedenborg Society, 2005.

Bergthaler, Hannes. "Limits of Agency: Notes on the Material Turn from a Systems-Theoretical Perspective." In *Material Ecocriticism*, edited by Serenella Iovino and Serpil Opperman, 37–50. Bloomington: Indiana University Press, 2014.

Berlin, Isaiah. *The Roots of Romanticism*. Princeton, N.J.: Princeton University Press, 2001.

Berry, Evan. *Devoted to Nature: The Religious Roots of American Environmentalism*. Oakland: University of California Press, 2015.

Blake, William. *The Complete Poetry and Prose of William Blake*. Edited by David V. Erdman and Harold Bloom. Berkeley: University of California Press, 1982.

———. *The Illuminated Books of William Blake*, Vol. 1: *Jerusalem: The Emanation of the Giant Albion*. Edited by Morton Paley and David Bindman. Princeton, N.J.: Princeton University Press, 1997.

Block, Marguerite Beck. *The New Church in the New World: A Study of Swedenborgianism in America*. New York: Octagon Books, 1968.

Blumenberg, Hans. *Die Lesbarkeit der Welt* [The readability of the world]. Berlin: Suhrkamp Verlag, 1986.

Borges, Jorge Luis. *Selected Non-Fictions*. Edited by Eliot Weinberger. Translated by Esther Allen, Suzanne Jill Levine, and Eliot Weinberger. New York: Penguin, 1999.

Bosco, Ronald A. "His Lectures were Poetry, His Teaching the Music of the Spheres: Annie Adams Fields and Francis Greenwood Peabody on Emerson's 'Natural

History of the Intellect' University Lectures at Harvard in 1870." *Harvard Library Bulletin* 8, no. 2 (1997): 3–79.

Bosley, Ted. "A. C. Schweinfurth." In *Toward a Simpler Way of Life: The Arts and Crafts Architects of California*, edited by Robert Winter, 11–22. Berkeley: University of California Press, 1997.

Brechin, Gray. *Imperial San Francisco: Urban Power, Earthly Ruin, with a New Preface.* Berkeley: University of California Press, 2006.

Brewster, Carrie, ed. *The Comprehensive Keith: The Hundred Year History of the Saint Mary's Collection of Works by William Keith.* With contributions from Alfred C. Harrison Jr., Jeanne McKee Rothe and Andrea Rothe, and Julian Rothe. Moraga: St. Mary's College of California, 2011.

Brideoake, Fiona. "Extraordinary Female Affection: The Ladies of Llangollen and the Endurance of Queer Community." *Romanticism on the Net* 36–37 (2004–5), https://www.erudit.org/en/journals/ron/2004-n36-37-ron947/011141ar/.

Brøgger, Fredrik. "The Conception of Nature in the American Romantic Age Is Anthropomorphic? Anthropocentric? Other?: Excursions into Emerson, Thoreau, and Melville." In *"Nature's Nation" Revisited: American Conceptions of Nature from Wonder to Ecological Crisis*, edited by Hans Bak and Walter W. Hölbling, 94–113. Amsterdam: VU University Press, 2003.

Brown, Bill. *Other Things.* Chicago: University of Chicago Press, 2015.

Brownlee, Peter John. *Manifest Destiny / Manifest Responsibility: Environmentalism and the Art of the American Landscape.* Loyola University Museum of Art. Chicago: Terra Foundation for American Art, 2008.

Brunner, Ernst. *Darra: Om Swedenborg.* [Trembling: On Swedenborg]. Stockholm: Bonnier, 2017.

Bubandt, Nils. "Anthropocene Uncanny: Nonsecular Approaches to Environmental Change." In *A Non-secular Anthropocene: Spirits, Specters, and Other Nonhumans in a Time of Environmental Change*, edited by Nils Bubandt, 2–18. Aarhus: AURA Working Papers Vol. 3, 2018.

———. "Haunted Geologies: Spirits, Stones, and the Necropolitics of the Anthropocene." In *Arts of Living on a Damaged Planet*, edited by Anna Tsing, Heather Swanson, Elaine Gan, and Nils Bubandt, G121–41. Minneapolis: University of Minnesota Press, 2017.

Budd, Louis J. "The American Background." In *The Cambridge Companion to American Realism and Naturalism: From Howells to London*, edited by Donald Pizer, 21–46. Cambridge: Cambridge University Press, 1995.

Buell, Lawrence. *Emerson.* Cambridge, Mass.: Belknap Press of Harvard University Press, 2003.

———. *The Environmental Imagination: Thoreau, Nature Writing, and the Formation of American Culture.* Cambridge, Mass.: Belknap Press of Harvard University Press, 1995.

———. *The Future of Environmental Criticism: Environmental Crisis and Literary Imagination.* Malden, Mass.: Blackwell, 2005.

———. *Writing for an Endangered World: Literature, Culture, and Environment in the U.S. and Beyond*. Cambridge, Mass.: Belknap Press of Harvard University Press, 2001.

Buhle, Paul, and Noah Van Sciver. *Johnny Appleseed: Green Spirit of the Frontier*. Seattle, Wash.: Fantagraphics Books, 2017.

Burton, Lloyd. *Worship and Wilderness: Culture, Religion, and Law in Public Lands Management*. Madison: University of Wisconsin Press, 2002.

Butler, Judith. *Gender Trouble: Feminism and the Subversion of Identity*. New York: Routledge, 2006.

Butterfield, Andrew. "The Genius of George Inness." *New York Review of Books*, September 15, 2008, 8–10.

Callicott, J. Baird, and Michael P. Nelson, eds. *The Great New Wilderness Debate*. Athens: University of Georgia Press, 1998.

———. *The Wilderness Debate Rages On: Continuing the Great New Wilderness Debate*. Athens: University of Georgia Press, 2008.

Cameron, Kenneth Walter. *Emerson the Essayist; an Outline of His Philosophical Development through 1836, with Special Emphasis on the Sources and Interpretation of Nature*. 2 vols. Raleigh, N.C.: Thistle Press, 1945.

Cameron, Sharon. *Impersonality: Seven Essays*. Chicago: University of Chicago Press, 2007.

———. "The Way of Life by Abandonment: Emerson's Impersonal." *Critical Inquiry* 25, no 1 (1988): 1–31.

Carlyle, Thomas. "T.C. to James John Garth Wilkinson." *The Carlyle Letters Online* 10, no. 1 (August 2, 1838): 141–42.

Carlyle, Thomas, and Jane Welsh Carlyle. *The Collected Letters of Thomas and Jane Welsh Carlyle*. Vols. 10–11. Edited by Ian Campbell, Aileen Christianson, and David R. Sorenson. Durham, N.C.: Duke University Press, 1976–77.

Carroll, Siobhan. "Crusades against Frost: Frankenstein, Polar Ice, and Climate Change in 1818." *European Romantic Review* 24, no. 2 (2013): 211–30.

Cartwright, Jacob, with Nick Jordan and Stephen McNeilly, directors. *Heaven, Hell, and Other Places: A Film about Emanuel Swedenborg*. With Philip Makatrewicz and Gary Lachman. DVD. London: Swedenborg Society, 2011.

Castoriadus, Cornelius. *The Imaginary Institution of Society*. Cambridge, Mass.: MIT Press, 1987.

Cavell, Stanley. *Conditions Handsome and Unhandsome: The Constitution of Emersonian Perfectionism*. Chicago: University of Chicago Press, 1991.

———. *Emerson's Transcendental Etudes*. Stanford, Calif.: Stanford University Press, 2003.

———. *The Senses of Walden*. Chicago: University of Chicago Press, 1992.

———. *This New Yet Unapproachable America: Lectures after Emerson after Wittgenstein*. Chicago: University of Chicago Press, 2013.

Chakrabarty, Dipesh. "The Climate of History: Four Theses." *Critical Inquiry* 35, no. 2 (2009): 197–222.

Chevigny, Bell Gale. *The Woman and the Myth: Margaret Fuller's Life and Writings.* Boston: Northeastern University Press, 1993.

Cikovsky, Nicolai. *George Inness.* New York: Harry N. Abrams, 1993.

———. "George Inness and the Hudson River School: 'The Lackawanna Valley.'" *American Art Journal* 2, no. 2 (1970): 36–57.

———. "'The Ravages of the Ax': The Meaning of the Tree Stump in Nineteenth-Century American Art." *Art Bulletin* 61 (1979): 611–26.

Cohen, Jeffrey Jerome. "Foreword: Storied Matter." In *Material Ecocriticism,* edited by Serenella Iovino and Serpil Opperman, ix–xii. Bloomington: Indiana University Press, 2014.

Cohen, Michael P. *The History of the Sierra Club 1892–1970.* San Francisco: Random House, 1988.

———. *The Pathless Way: John Muir and American Wilderness.* Madison: University of Wisconsin Press, 1986.

Cole, Thomas. "Essay on American Scenery." *American Monthly Magazine* 1 (January 1836): 1–12.

Copa, Adrienne. "The Garden Church of San Francisco: A History." Unpublished manuscript, n.d. Archives of the San Francisco Swedenborgian Church.

Congressional Record Containing the Proceedings and Debates of the Sixty-third Congress, First Session. Volume 50, Part 6, 5441–77. Washington, D.C.: Proceedings of Congress and Congressional Publications, 1913.

Corey, Joshua, and G. C. Waldrep, eds. *The Arcadia Project: North American Postmodern Pastoral.* Boise, Idaho: Ahsahta Press, 2012.

Corn, Wanda. "The Color of Mood: American Tonalism, 1880–1910 (1972)." In *Like Breath on Glass: Whistler, Inness, and the Art of Painting Softly,* edited by Marc Simpson, 228–47.

Cornelius, Brother Fidelis. *Keith: Old Master of California.* New York: G. P. Putnam's Sons, 1942.

———. "William Keith, Rembrandt of Landscape." *Moraga Quarterly* 1, no. 2 (1930): 65–77.

Crapanzano, Vincent. *Imaginative Horizons: An Essay in Literary-Philosophical Anthropology.* Chicago: University of Chicago Press, 2004.

Crary, Jonathan. *Techniques of the Observer: On Vision and Modernity in the 19th Century.* Cambridge, Mass.: MIT Press, 1992.

Croce, Paul Jerome. *Science and Religion in the Era of William James: Eclipse of Certainty, 1820–1880.* Chapel Hill: University of North Carolina Press, 1995.

———. *Young William James Thinking.* Baltimore: Johns Hopkins University Press, 2017.

Cronon, William. *Uncommon Ground: Toward Reinventing Nature.* New York: Norton, 1995.

———, ed. *Uncommon Ground: Rethinking the Human Place in Nature.* New York: Norton, 1996.

Crowley, Karlyn. "'One [Hermaphroditic] Angel': Swedenborg, Gender Complementarity, and Divine Love in Julia Ward Howe's 'The Hermaphrodite.'" In *Nineteenth-Century American Women Write Religion: Lived Theologies and Literature,* edited by Mary McCartin Wearn, 89–104. Burlington, Vt.: Ashgate Publishing, 2014.

Crutzen, Paul, and Eugene F. Stormer. "The Anthropocene." *IGBP Newsletter* 41 (May 2000): 17–18.

Dann, Kevin. *Expect Great Things: The Life and Search of Henry David Thoreau.* New York: TarcherPerigee, 2018.

Davenport, Coral, and Campbell Robertson. "Resettling the First American 'Climate Refugees.'" *New York Times,* May 3, 2016. Accessed July 18, 2016. http://www.nytimes.com/2016/05/03/us/resettling-the-first-american-climate-refugees.html.

Debord, Guy. "Introduction to a Critique of Urban Geography." Translated by Ken Knabb. In *Critical Geographies: A Collection of Readings,* edited by Harald Bauder and Salvatore Engel-Di Mauro, 23–27. Kelowna, B.C.: Praxis (e)Press, 2008.

DeConto, Robert M., and David Pollard. "Contribution of Antarctica to Past and Future Sea-Level Rise." *Nature* 531(March 31, 2016): 591–97.

Deleuze, Gilles. *Cinema 1: The Movement-Image.* Translated by Hugh Tomlinson and Barbara Habberjam. Minneapolis: University of Minnesota Press, 1986.

———. *Cinema 2: The Time-Image.* Translated by Hugh Tomlinson and Robert Galeta. Minneapolis: University of Minnesota Press, 1989.

———. *Difference and Repetition.* Translated by Paul Patton. New York: Columbia University Press, 1994.

———. *The Logic of Sense.* Translated by Mark Lester. London: Athlone Press, 1990.

Deleuze, Gilles, and Félix Guattari. *A Thousand Plateaus: Capitalism and Schizophrenia.* Translated by Brian Massumi. Minneapolis: University of Minnesota Press, 1987.

DeLue, Rachel Ziady. *George Inness and the Science of Landscape.* Chicago: University of Chicago Press, 2004.

Derrida, Jacques. *Of Grammatology.* Translated by Gayatri Chakravorty Spivak. 2nd ed. Baltimore: Johns Hopkins University Press, 2016.

———. *Specters of Marx: The State of Debt, the Work of Mourning, and the New International.* Translated by Peggy Kamuf. New York: Routledge, 1994.

———. *Writing and Difference.* Translated by Alan Bass. Chicago: University of Chicago Press, 1978.

Dole, George F. "Philosemitism in the 17th Century." *Studia Swedenborgiana* 7, no. 1 (December 1990): 5–17.

Dole, George F., and Robert H. Kirven. *A Scientist Explores Spirit: A Compact Biography of Emanuel Swedenborg with Key Concepts of Swedenborg's Theology.* New York: Swedenborg Foundation, 1992.

Donovan, Josephine. "Jewett and Swedenborg." *American Literature* 65, no. 4 (1993): 731–51.

———. *Sarah Orne Jewett.* New York: Ungar, 1980.

Dumas, Jean-Baptiste, and M. (Amand) Bineau. *Leçons sur la philosophie chimique* [Lessons on the philosophy of chemistry]. Brussels: Société Belge de Librairie, 1839.

Dunaway, Finis. *Natural Visions: The Power of Images in American Environmental Reform*. Chicago: University of Chicago Press, 2005.

———. *Seeing Green: The Use and Abuse of American Environmental Images*. Chicago: University of Chicago Press, 2015.

Dunér, David. *The Natural Philosophy of Emanuel Swedenborg: A Study in the Conceptual Metaphors of the Mechanistic World-View*. Translated by Alan Crozier. London: Springer, 2012.

ecopoetics. Edited by Jonathan Skinner. 7 vols. Buffalo, N.Y.: Periplum Editions, 2001–9.

Ek, Anna, with Stefan Löfgren and Johan Bergholm. "Environmental Effects of One Thousand Years of Copper Production at Falun, Central Sweden." *Ambio* 30, no. 2 (2001): 96–103.

Emerson, Ralph Waldo. *"The Best Read Naturalist": Nature Writings of Ralph Waldo Emerson*. Edited by Michael Branch and Clinton Mohs. Charlottesville: University of Virginia Press, 2017.

———. *The Complete Works of Ralph Waldo Emerson*, Vol. 8: *Letters and Social Aims*. New York: Houghton and Mifflin, 1904.

———. *The Correspondence of Emerson and Carlyle*. New York: Columbia University Press, 1964.

———. *Essays and Lectures*. New York: Library of America, 1983.

———. *Essays and Poems*. New York: Library of America, 1996.

———. *Journals and Miscellaneous Notebooks*. 16 vols. Cambridge, Mass.: Harvard University Press, 1960–82.

———. *The Later Lectures of Ralph Waldo Emerson, 1843–1871*. Edited by Ronald A. Bosco and Joel Myerson. 2 vols. Athens: University of Georgia Press, 2001.

———. *The Letters of Ralph Waldo Emerson*. Edited by Ralph Rusk. 10 vols. New York: Columbia University Press, 1941.

Emmett, Robert, and Thomas Lekan. *Whose Anthropocene? Revisiting Dipesh Chakrabarty's 'Four Theses.'* RCC Perspectives Series. Munich: Rachel Carson Center, 2016.

Erickson, Jacob J. "Theophanic Materiality: Political Ecology, Inhuman Touch, and the Art of Andy Goldsworthy." In *Entangled Worlds: Religion, Science, and New Materialisms*, edited by Catherine Keller and Mary-Jane Rubenstein, 203–20. New York: Fordham University Press, 2017.

Esterson, Rebecca. "Secrets of Heaven: Allegory, the Jews, and the Case of Emanuel Swedenborg." PhD diss., Boston University, 2017.

Fetterley, Judith, and Marjorie Pryse. *Writing out of Place: Regionalism, Women, and American Literary Culture*. Urbana: University of Illinois Press, 2005.

Finger, Stanley. *Origins of Neuroscience: A History of Explorations in Brain Function*. New York: Oxford University Press, 2001.

Fisher, Irving. "An Iconology of City Planning: The Plan of Chicago." In *Swedenborg and His Influence,* edited by Erland J. Brock, 449–64. Bryn Athyn, Pa.: Academy of the New Church, 1988.

Fleck, Richard F. *Henry Thoreau and John Muir among the Native Americans.* Portland, Ore.: West Winds Press, 2015.

Fletcher, Angus. *A New Theory for American Poetry: Democracy, the Environment, and the Future of Imagination.* Cambridge, Mass.: Harvard University Press, 2004.

Foote-Smith, E., and T. J. Smith. "Emanuel Swedenborg." *Epilepsia* 37, no. 2 (February 1996): 211–18.

Fors, Hjalmar. *The Limits of Matter: Chemistry, Mining, and the Enlightenment.* Chicago: University of Chicago Press, 2015.

Forster, Wolfgang. *Karl Christian Friedrich Krauses frühe Rechtsphilosophie und ihr geistesgeschichtlicher Hintergrund* [Karl Christian Friedrich Krause's early legal philosophy and its place in the history of ideas]. Ebelsbach: Aktiv Druck & Verlag, 2000.

Foucault, Michel. *Archeology of Knowledge and the Discourse on Language.* Translated by A. M. Sheridan Smith. New York: Pantheon Books, 1972.

———. *Madness and Civilization: A History of Insanity in the Age of Reason.* Translated by Richard Howard. New York: Vintage Books, 1988.

Fox, Stephen R. *The American Conservation Movement: John Muir and His Legacy.* Madison: University of Wisconsin Press, 1981.

Francis, Pope. *Encyclical Letter Laudato Si of the Holy Father Francis: On Care for Our Common Home.* Encyclical. Vatican City: Vatican Press, 2015.

Freeman, Elizabeth. *Time Binds: Queer Temporalities, Queer Histories.* Durham, N.C.: Duke University Press, 2010.

Freudenheim, Leslie Mandelson. *Building with Nature: Inspiration for the Arts & Crafts Home.* Salt Lake City: Gibbs Smith, 2005.

Friesner, Nicholas Aaron. "A Transcendentalist Nature Religion." *Religions* 8, no. 8 (2017): 1–18. Accessed April 15, 2018. http://www.mdpi.com/207-1444/8/8/130.

Fukayama, Francis. *The End of History and the Last Man.* New York: Free Press Books, 1982.

———. "Is American Democracy Strong Enough for Trump? The Case against Panic." *Politico Magazine,* January 23, 2017. Accessed December 30, 2018. https://www.politico.com/magazine/story/2017/01/donald-trump-american-democracy-214683.

Gaard, Greta. "Mindful New Materialisms: Buddhist Roots for Material Ecocriticism's Flourishing." In *Material Ecocriticism,* edited by Serenella Iovino and Serpil Opperman, 291–300. Bloomington: Indiana University Press, 2014.

Gabay, Alfred. *The Covert Enlightenment: Eighteenth-Century Counterculture and Its Aftermath.* West Chester, Pa.: Swedenborg Foundation, 2005.

Gatta, John. *Making Nature Sacred: Literature, Religion, and Environment in America from the Puritans to the Present.* New York: Oxford University Press, 2004.

Gerhardt, Christine. *A Place for Humility: Whitman, Dickinson, and the Natural World.* Iowa City: University of Iowa Press, 2014.

Ghosh, Amitav. *The Great Derangement: Climate Change and the Unthinkable.* Chicago: University of Chicago Press, 2017.

Gigante, Denise. *Life: Organic Form and Romanticism.* New Haven, Conn.: Yale University Press, 2009.

Gladish, Robert. *Swedenborg, Fourier, and the America of the 1840s.* Bryn Athyn, Pa.: Swedenborg Scientific Association, 1982.

Graham, Frank, and Carl W. Buchheister. *The Audubon Ark: A History of the National Audubon Society.* Austin: University of Texas Press, 1992.

Grant, Iain Hamilton. *Philosophies of Nature after Schelling.* London: Continuum, 2006.

Green, Tyler. *Carlton Watkins: Making the West American.* Berkeley: University of California Press, 2018.

Gross, Charles G. *Brain, Vision, Memory: Tales in the History of Neuroscience.* Cambridge, Mass.: MIT Press, 1998.

Gura, Philip F. *The Wisdom of Words: Language, Theology, and Literature in the New England Renaissance.* Middletown, Conn.: Wesleyan University Press, 1981.

Halberstam, Judith/Jack. *In a Queer Time and Place: Transgender Bodies, Subcultural Lives.* New York: NYU Press, 2005.

Hallengren, Anders. *The Code of Concord: Emerson's Search for Universal Laws.* Stockholm: Almqvist and Wiksel, 1994.

———. *Gallery of Mirrors: Reflections of Swedenborgian Thought.* West Chester, Pa.: Swedenborg Foundation, 1998.

———. *The Grand Theme: And Other Essays.* London: Swedenborg Society, 2013.

———. "A Hermeneutic Key to the Title 'Leaves of Grass.'" In *In Search of the Absolute: Essays on Swedenborg and Literature,* edited by Stephen McNeilly, 45–59. London: Swedenborg Society, 2004.

Haller, John. *Distant Voices: Sketches of a Swedenborgian World View.* London: Swedenborg Society, 2017.

———. *The History of New Thought: From Mind Cure to Positive Thinking and the Prosperity Gospel.* West Chester, Pa.: Swedenborg Foundation Publishers, 2012.

———. *Swedenborg, Mesmer, and the Mind/Body Connection: The Roots of Complementary Medicine.* West Chester, Pa.: Swedenborg Foundation, 2010.

Hallowell, A. Irving. "Ojibwa Ontology, Behavior, and World-View." In *Teachings from the American Earth,* edited by Dennis and Barbara Tedlock, 141–79. New York: Liveright, 1975.

Hanegraaff, Wouter J. "Emanuel Swedenborg, the Jews, and Jewish Traditions." In *Reuchlin und seine Erben: Forscher, Denker, Ideologen, und Spinner* [Reuchlin and his heirs: Researchers, thinkers, ideologues, and fabulists], edited by Peter Schäfer, 135–54. Ostfildern: Jan Thorbecke, 2004.

———. *New Age Religion and Western Culture: Esotericism in the Mirror of Secular Thought.* Albany: State University of New York Press, 1998.

———. “Swedenborg aus der Sicht von Kant und der akademischen Kantforschung” [Swedenborg from the perspective of Kant and academic research on Kant]. In *Kant und Swedenborg: Zugänge zu einem umstrittenen Verhältnis* [Kant and Swedenborg: Different approaches to a difficult relationship], edited by Friedemann Stengel, 157–72. Tübingen: Max Niemeyer Verlag, 2008.

Haraway, Donna J. *Crystals, Fabrics, and Fields: Metaphors That Shape Embryos.* Berkeley, Calif.: North Atlantic Books, 2004.

———. *Staying with the Trouble: Making Kin in the Chthulucene.* Durham, N.C.: Duke University Press, 2016.

Harding, Walter. *Emerson's Library.* Charlottesville: University Press of Virginia, 1967.

———. “Swedenborgian Spirit and Thoreauvian Sense: Another Look at Correspondence.” *Journal of American Studies* 8, no. 1 (1974): 65–79.

Harding, Walter, and Michael Meyer. *The New Thoreau Handbook.* New York: NYU Press, 1980.

Harman, Graham. *The Third Table: 100 Notes, 100 Thoughts.* Documenta Series 085. Ostfildern: Hatje Cantz, 2012.

Harrison, Alfred C., Jr. “The Art of William Keith.” In *The Comprehensive Keith: The Hundred Year History of the Saint Mary's College Collection of Works by William Keith,* edited by Carrie Brewster, 3–87. Moraga: St. Mary's College of California, 2011.

Hartley, George Inness. *The Importance of Bird Life: A Popular Account of Its Economic Significance and Conservation.* New York: Century, 1922.

———. *The Last Parrakeet.* Illustrated by Donald Teague. New York: Century, 1923.

Harvard College Library. “A List of the Books from the Bequest of Theodore Jewett Eastman That Bear the Marks of Ownership of Sarah Orne Jewett” (1936). Harvard University Archives, UAIII 50.29.33.10.

Harvey, Graham. *Animism: Respecting the Living World.* New York: Columbia University Press, 2005.

Hawthorne, Nathaniel. *Tales and Sketches.* New York: Library of America, 1996.

Heinrichs, Michael. *Emanuel Swedenborg in Deutschland: Eine kritische Darstellung der Rezeption des schwedischen Visionärs im 18. und 19. Jahrhundert* [Emanuel Swedenborg in Germany: A critical study of the reception of the Swedish visionary in the 18th and 19th centuries]. Berlin: Peter Lang, 1979.

Heise, Ursula K. *Imagining Extinction: The Cultural Meanings of Endangered Species.* Chicago: University of Chicago Press, 2016.

———. *Sense of Place and Sense of Planet: The Environmental Imagination of the Global.* New York: Oxford University Press, 2008.

Henry, Daniel Lee. *Across the Shaman's River: John Muir, the Tlingit Stronghold, and the Opening of the North.* Fairbanks: University of Alaska Press, 2017.

Herzogenrath, Bernd. “The ‘Weather of Music’: Sounding Nature in the Twentieth and Twenty-first Centuries.” In *Deleuze / Guattari and Ecology,* edited by Bernd Herzogenrath, 216–32. London: Palgrave Macmillan, 2009.

———. "White." In *Prismatic Ecology: Ecotheory beyond Green*, edited by Jeffrey Jerome Cohen, 1–21. Minneapolis: University of Minnesota Press, 2013.

Hitchcock, Ethan Allen. *Swedenborg, a Hermetic Philosopher: Being a Sequel to Remarks on Alchemy and the Alchemists. Showing That Emanuel Swedenborg Was a Hermetic Philosopher and That His Writings May Be Interpreted from the Point of View of Hermetic Philosophy*. New York: D. Appleton & Company, 1858.

Holmes, Richard. *The Age of Wonder: How the Romantic Generation Discovered the Beauty and Terror of Science*. New York: Pantheon Books, 2008.

Horn, Friedemann. *Schelling und Swedenborg: Ein Beitrag zur Problemgeschichte des Deutschen Idealismus und zur Geschichte Swedenborgs in Deutschland. Nebst Einem Anhang über K.C.F. Krause und Swedenborg* [Schelling and Swedenborg: A contribution to understanding a historical problem in German Idealism and the history of Swedenborg in Germany. With additional materials concerning K.C.F. Krause and Swedenborg]. Zürich: Swedenborg Verlag, 1954.

Hotson, Clarence Paul. "Emerson and Swedenborg." PhD diss., Harvard University, 1929.

———. "Emerson's Biographical Sources for 'Swedenborg.'" *Studies in Philology* 26, no. 1 (January 1929): 23–46.

Howe, Julia Ward. *The Hermaphrodite*. Edited by Gary Williams. Lincoln: University of Nebraska Press, 2009.

Howe, Nicolas. *Landscapes of the Secular: Law, Religion, and American Sacred Space*. Chicago: University of Chicago Press, 2016.

Hume, Angela, and Gillian Osborne, eds. *Ecopoetics: Essays in the Field*. Iowa City: University of Iowa Press, 2018.

Hutchings, Kevin. *Imagining Nature: Blake's Environmental Poetics*. Toronto: McGill-Queens University Press, 2003.

Iijima, Brenda, ed. *Eco Language Reader*. Brooklyn, N.Y.: Portable Press at Yo-Yo Labs, 2010.

Inness, George. *George Inness: Writings and Reflections on Art and Philosophy*. Edited by Adrienne Baxter Bell. New York: George Braziller, 2007.

Iovino, Serenella. "Behold." In *Veer Ecology: A Companion for Environmental Thinking*, edited by Jeffrey Jerome Cohen and Lowell Duckert, 312–26. Minneapolis: University of Minnesota Press, 2017.

IPCC [Intergovernmental Panel on Climate Change]. "Global Warming of 1.5° Celsius: An IPCC Special Report on the Impacts of Global Warming of 1.5 above Pre-industrial Levels and Related Global Greenhouse Gas Emission Pathways, in the Context of Strengthening the Global Response to the Threat of Climate Change, Sustainable Development, and Efforts to Eradicate Poverty." Edited by V. Masson-Delmotte, P. Zhai, H. O. Pörtner, D. Roberts, J. Skea, P. R. Shukla, A. Pirani, W. Moufouma-Okia, C. Péan, R. Pidcock, S. Connors, J.B.R. Matthews, Y. Chen, X. Zhou, M. I. Gomis, E. Lonnoy, T. Maycock, M. Tignor, and T. Waterfield. In Press. Accessed November 3, 2018. http://www.ipcc.ch/report/sr15/.

James, William. *Writings, 1878–1899: Psychology, Briefer Course / The Will to Believe / Talks to Teachers and Students / Essays.* New York: Library of America, 1992.

———. *Writings 1902–1910: The Varieties of Religious Experience / Pragmatism / A Pluralistic Universe / The Meaning of Truth / Some Problems of Philosophy / Essays.* New York: Library of America, 1988.

Jewett, Sarah Orne. Correspondence to/from Theophilus Parsons. *The Sarah Orne Jewett Text Project.* Accessed December 18, 2013. http://www.sarahornejewett.org/soj/contents.htm.

———. *Novels and Stories: Deephaven; A Country Doctor; The Country of the Pointed Firs; Stories & Sketches.* New York: Library of America, 1996.

Johnson, Gregory. "Introduction: Kant's Strangest Book." In *Kant on Swedenborg: Dreams of a Spirit-Seer and Other Writings,* edited by Gregory Johnson, xi–xxxvi. West Chester, Pa.: Swedenborg Foundation, 2002.

Jonsson, Inge. *A Drama of Creation: Sources and Influences in Swedenborg's Worship and Love of God.* Translated by Matilda McCarthy. West Chester, Pa.: Swedenborg Foundation, 2004.

———. *Visionary Scientist: The Effects of Science and Philosophy on Swedenborg's Cosmography.* West Chester, Pa.: Swedenborg Foundation, 1999.

Kant, Immanuel. *Kant on Swedenborg: Dreams of a Spirit-Seer and Other Writings.* Edited by Gregory Johnson. Translated by Gregory Johnson and Glenn Alexander Magee. West Chester, Pa.: Swedenborg Foundation, 2002.

Kearney, Richard. "A Dialogue with Jacques Derrida." *Philosophy Today* 48, no. 1 (2004): 4–11.

———. *Dialogues with Contemporary Continental Thinkers: Paul Ricoeur, Emmanuel Levinas, Herbert Marcuse, Stanislas Breton, Jacques Derrida.* Manchester: Manchester University Press, 1984.

Keith, William. Keith-McHenry-Pond Papers, 1841–1961. BANC MSS C-B 595. Box 1—Letters Written by Keith. Bancroft Library, University of California, Berkeley.

———. "Lecture by William Keith, 1894 or 1895." Inez Henderson Pond Collection, 1827–1962. Mss10. Series III, Box 1. Holt-Atherton Special Collections, University of the Pacific.

Keller, Catherine. *Cloud of the Impossible: Negative Theology and Planetary Entanglement.* New York: Columbia University Press, 2014.

Kermode, Frank. *The Sense of an Ending: Studies in the Theory of Fiction.* New York: Oxford University Press, 2000.

Kerrigan, William. *Johnny Appleseed and the American Orchard: A Cultural History.* Baltimore: Johns Hopkins University Press, 2012.

Klein, Naomi. *This Changes Everything: Capitalism vs. the Climate.* New York: Simon and Schuster, 2015.

Knapp, Horace S. *A History of the Pioneer and Modern Times of Ashland County, from the Earliest to the Present Date.* Philadelphia: J. B. Lippincott, 1863.

Kolbert, Elizabeth. *The Sixth Extinction: An Unnatural History.* New York: Picador, 2015.

Kosky, Jeffrey L. *Arts of Wonder: Enchanting Secularity: Walter de Maria, Diller + Scofidio, James Turrell, Andy Goldsworthy.* Chicago: University of Chicago Press, 2013.

Kroeber, Karl. *Ecological Literary Criticism: Romantic Imagining and the Biology of Mind.* New York: Columbia University Press, 1994.

Lachman, Gary. *Swedenborg: An Introduction to His Life and Ideas.* New York: Tarcher-Perigee, 2012.

Lakoff, George, and Mark Johnson. *Metaphors We Live By.* Chicago: University of Chicago Press, 2003.

Latour, Bruno. "An Attempt at a Compositionist Manifesto." *New Literary History* 41, no. 3 (Summer 2010): 471–90.

———. *Facing Gaia: Eight Lectures on the New Climactic Regime.* Translated by Catherine Porter. Medford, Mass.: Polity Press, 2017.

———. *We Have Never Been Modern.* Translated by Catherine Porter. Cambridge, Mass.: Harvard University Press, 1993.

Lawrence, James. "Correspondentia: A Neologism by Aquinas Attains Its Zenith in Swedenborg." *Correspondences* 5 (2017): 41–63.

Lears, T. J. Jackson. *No Place of Grace: Antimodernism and the Transformation of American Culture, 1880–1920.* Chicago: University of Chicago Press, 1981.

Limbaugh, Ronald H. "California's Kindred Spirits: John Muir and William Keith." In *John Muir: Family, Friends, Adventures,* edited by Sally Miller and Daniel Morrison, 65–80. Albuquerque: University of New Mexico Press, 2005.

———. "John Muir and the Mining Industry." *Mining History Journal* 3 (1996): 61–66.

———. *John Muir's "Stickeen" and the Lessons of Nature.* Fairbanks: University of Alaska Press, 1996.

Lindqvist, Sante. *Technology on Trial: the Introduction of Steam Power Technology into Sweden, 1715–1736.* Stockholm: Almqvist and Wiksell International, 1984.

Lindroth, Sten. *Swedish Men of Science, 1650–1950.* Stockholm: Swedish Institute, 1952.

Lindsay, Vachel. *Johnny Appleseed and Other Poems.* New York: Macmillan, 1928.

———. *The Village Magazine.* Springfield, Ill.: Jefferson Printing Company, 1920.

London, Charmian. *The Book of Jack London.* 2 vols. New York: Century, 1921.

Lowenthal, David, and William Cronon. *George Perkins Marsh: Prophet of Conservation.* Seattle: University of Washington Press, 2003.

Luciano, Dana. "Sacred Theories of Earth: Matters of Spirit in the Soul of Things." *American Literature* 86, no. 4 (December 1, 2014): 713–36.

Maran, Timo. "Where Do Your Borders Lie? Reflections on the Semiotical Ethics of Nature." In *Nature in Literary and Cultural Studies: Transatlantic Conversations on Ecocriticism,* edited by Catrin Gersdorf and Sylvia Mayer, 455–76. Amsterdam: Rodopi, 2006.

Marsh, George Perkins. *Man and Nature; or, Physical Geography as Modified by Human Action.* New York: Charles Scribner, 1864.

Marx, Leo. *The Machine in the Garden: Technology and the Pastoral Ideal in America.* Reprint. New York: Oxford University Press, 2000.

Matthews, Freya. *For Love of Matter: A Contemporary Panpsychism.* Albany: State University of New York Press, 2003.

Matthiessen, Francis Otto. *American Renaissance: Art and Expression in the Age of Emerson and Whitman.* New York: Oxford University Press, 1964.

———. *Sarah Orne Jewett.* With Illustrations by Russell Cheney. Boston: Houghton Mifflin, 1929.

Mayer, Jean-François. "War Swedenborg der Herold des New Age?" [Was Swedenborg the herald of the new age?] *Offene-Tore* 42, no. 4 (1998): 186–99.

Mayer, Sylvia. *Naturethik und Neuengland-Regionalliteratur: Harriet Beecher Stowe, Rose Terry Cooke, Sarah Orne Jewett, Mary E. Wilkins Freeman* [The ethics of nature and New England literature: Harriet Beecher Stowe, Rose Terry Cooke, Sarah Orne Jewett, Mary E. Wilkins Freeman]. Heidelberg: Universitätsverlag Winter, 2004.

Mazow, Leo. "George Inness, Henry George, the Single Tax, and the Future Poet." *American Art* 18 (Spring 2004): 59–77.

———. "George Inness: Problems in Antimodernism." Ph.D. diss., University of North Carolina, Chapel Hill, 1996.

McKusick, James C. *Green Writing: Romanticism and Ecology.* New York: St. Martin's Press, 2000.

McMurry, Andrew. *Environmental Renaissance: Emerson, Thoreau, and the American System of Nature.* Athens: University of Georgia Press, 2003.

Meillassoux, Quentin. *After Finitude: An Essay on the Necessity of Contingency.* Translated by Ray Brassier. London: Continuum, 2009.

Merchant, Carolyn. *Spare the Birds!: George Bird Grinnell and the First Audubon Society.* New Haven, Conn.: Yale University Press, 2016.

Miller, Angela. "The Fate of Wilderness in American Landscape Art." In *A Keener Perception: Ecocritical Studies in American Art History,* edited by Alan C. Braddock and Christoph Irmscher, 85–109. Tuscaloosa: University of Alabama Press, 2009.

Miller, Perry. *Errand into the Wilderness.* Cambridge, Mass.: Harvard University Press, 1984.

———. *Nature's Nation.* Cambridge, Mass.: Belknap Press of Harvard University Press, 1967.

———. *The New England Mind, from Colony to Province.* Cambridge, Mass.: Belknap Press of Harvard University Press, 1983.

———. *The New England Mind: The Seventeenth Century.* Cambridge, Mass.: Harvard University Press, 1983.

———, ed. *The Transcendentalists: An Anthology.* Cambridge, Mass.: Harvard University Press, 1978.

Miłosz, Czesław. *Miłosz's ABCs.* New York: Farrar, Straus and Giroux, 2001.

Milton, John. *Paradise Lost.* London: Penguin Books, 2003.

Mitchell, W. J. T. "Imperial Landscape." In *Landscape and Power,* edited by W. J. T. Mitchell, 5–33. Chicago: University of Chicago Press, 2002.

Mooney, Chris. "Earth's Atmosphere Just Crossed Another Troubling Climate Change Threshold." *Washington Post,* May 3, 2018.

Moore, Charles. *Daniel H. Burnham: Architect, Planner of Cities.* 2 vols. Boston: Houghton Mifflin, 1921.

Morton, Timothy. *Being Ecological.* Cambridge, Mass.: MIT Press, 2018.

———. *Dark Ecology: For a Logic of Future Coexistence.* New York: Columbia University Press, 2016.

———. *The Ecological Thought.* Cambridge, Mass.: Harvard University Press, 2010.

———. *Ecology without Nature: Rethinking Environmental Aesthetics.* Cambridge, Mass.: Harvard University Press, 2007.

———. "*Frankenstein* and Ecocriticism." In *The Cambridge Companion to Frankenstein,* edited by Andrew Smith, 143–57. Cambridge: Cambridge University Press, 2016.

———. "X-Ray." In *Prismatic Ecology: Ecotheory beyond Green,* edited by Jeffrey Jerome Cohen Cohen. 311–27. Minneapolis: University of Michigan Press, 2013.

Muir, John. *Nature Writings: The Story of My Boyhood and Youth; My First Summer in the Sierra; The Mountains of California; Stickeen; Essays.* New York: Library of America, 1997.

———. *Travels in Alaska.* Edited by Marion Randall Parsons. New York: Houghton Mifflin, 1915.

———. *The Yosemite.* New York: Century, 1911.

Muir, John, and Jeanne Carr. *Kindred and Related Spirits: The Letters of John Muir and Jeanne E. Carr.* Edited by Bonnie Johanna Gisel. Salt Lake City: University of Utah Press, 2001.

Murphy, Patrick D. "Grounding Anotherness and Answerability through Allonational Ecoliterature Formations." In *Nature in Literary and Cultural Studies: Transatlantic Conversations on Ecocriticism,* edited by Catrin Gersdorf and Sylvia Mayer, 417–34. Amsterdam: Rodopi, 2006.

Nash, Roderick Frazier. *The Rights of Nature: A History of Environmental Ethics.* Madison: University of Wisconsin Press, 1989.

———. *Wilderness and the American Mind.* New Haven, Conn.: Yale University Press, 2001.

Neuhaus, Eugen. *William Keith: The Man and the Artist.* Berkeley: University of California Press, 1938.

Nixon, Rob. *Slow Violence and the Environmentalism of the Poor.* Cambridge, Mass.: Harvard University Press, 2011.

Novak, Barbara. *Nature and Culture: American Landscape and Painting, 1825–1875.* New York: Oxford University Press, 2007.

Odhner, Carl Theophilus. *James John Garth Wilkinson: A Biographical Sketch.* Huntingdon Valley, Pa.: Academy Book Room, 1901.

Orange, Donna M. *Peirce's Conception of God: A Developmental Study.* Lubbock, Tex.: Institute for Studies in Pragmaticism, 1984.

Orr, Oliver H. *Saving American Birds: T. Gilbert Pearson and the Founding of the Audubon Movement.* Gainesville: University Press of Florida, 1992.

Osborne, Gillian. "Bad Seeds: Inhuman Poetics in Nineteenth-Century America." PhD diss., University of California, Berkeley, 2014.

Otto, Rudolf. *The Idea of the Holy: an Inquiry into the Non-Rational Factor in the Idea of the Divine and Its Relation to the Rational.* Translated by John W. Harvey. London: Oxford University Press, 1936.

Paley, Morton. "'A New Heaven Is Begun': Blake and Swedenborgianism." *Blake: An Illustrated Quarterly* 12, no. 2 (1979): 64–90.

Parsons, Theophilus. *Essays: First Series.* Boston: William Carter and Brother, 1862.

———. *Essays: Second Series.* Boston: Crosby, Nichols, and Company, 1856.

———. *The Infinite and the Finite.* Boston: Roberts Brothers, 1872.

———. *Outlines of the Religion and Philosophy of Swedenborg.* New York: New Church Press, 1876.

Pauley, Steve. "The Importance of John Muir's First Public Lecture, Sacramento, 1876." *John Muir Newsletter* 9, no. 1 (1998): 1–4.

Peet, Malcolm. *Medicine, Mysticism, and Mythology: Swedenborg and Nineteenth-Century Esoteric Culture.* London: Swedenborg Society 2018.

Peirce, Charles Saunders. "The Basis of Pragmaticism in the Normative Sciences." In *The Essential Peirce: Selected Philosophical Writings,* Vol. 2: *(1893–1913),* 371–97. Bloomington: Indiana University Press, 1998.

———. *The Collected Papers of Charles Saunders Peirce,* Vol. 8: *Reviews, Correspondence, and Bibliography.* Edited by Paul Weiss. Cambridge, Mass.: Harvard University Press, 1958.

Philippon, Daniel J. *Conserving Words: How American Nature Writers Shaped the Environmental Movement.* Athens: University of Georgia Press, 2005.

Pike, Sarah M. *For the Wild: Ritual and Commitment in Radical Eco-Activism.* Oakland: University of California Press, 2017.

Pipkin, Whitney. "How to Farm a Forest—and Feed a Neighborhood." *National Geographic,* October 31, 2016. Accessed April 12, 2019. https://www.nationalgeographic.com/people-and-culture/food/the-plate/2016/october/forest-garden-feast-features-gluten-free-acorn-bread/.

Poe, Edgar Allan. *Eureka.* Edited by Susan F. Levine. Urbana: University of Illinois Press, 2004.

Poirier, Richard. *Poetry and Pragmatism.* Cambridge, Mass.: Harvard University Press, 1992.

Pollan, Michael. *The Botany of Desire: A Plant's-Eye View of the World.* New York: Random House, 2002.

Porte, Joel. *Emerson and Thoreau: Transcendentalists in Conflict.* Middletown, Conn: Wesleyan University Press, 1966.

Powers, Richard. *The Overstory*. London: William Heinemann, 2018.

Price, Robert. *Johnny Appleseed: Man and Myth*. Bloomington: Indiana University Press, 1954.

Promey, Sally M. "The Ribband of Faith: George Inness, Color Theory, and the Swedenborgian Church." *American Art Journal* 26, no. 1/2 (1994): 45–65.

Quick, Michael. *George Inness: A Catalogue Raisonné*. 2 vols. New Brunswick, N.J.: Rutgers University Press, 2007.

Raby, Megan. *American Tropics: The Caribbean Roots of Biodiversity Science*. Chapel Hill: University of North Carolina Press, 2017.

Rauer, Constantin. *Wahn und Wahrheit: Kants Auseinandersetzung mit dem Irrationalen* [Delusion and truth: Kant's confrontation with the irrational]. Berlin: De Gruyter, 2012.

Reed, Sampson. *Sampson Reed: Primary Source Material for Emerson Studies*. Edited by George Dole and Sylvia Shaw. New York: Swedenborg Foundation, 1992.

Reynolds, David. *Walt Whitman*. New York: Oxford University Press, 2005.

Richards, Robert J. *The Romantic Conception of Life: Science and Philosophy in the Age of Goethe*. Chicago: University of Chicago Press, 2002.

Richardson, Joan. *A Natural History of Pragmatism: The Fact of Feeling from Jonathan Edwards to Gertrude Stein*. New York: Cambridge University Press, 2007.

———. *Pragmatism and American Experience: An Introduction*. New York: Cambridge University Press, 2014.

Richardson, Robert. *Emerson: The Mind on Fire*. Berkeley: University of California Press, 1995.

———. *Thoreau: A Life of the Mind*. Berkeley: University of California Press, 1988.

Rigby, Catherine. "Spirits That Matter: Pathways toward a Rematerialization of Religion and Spirituality." In *Material Ecocriticism*, edited by Serenella Iovino and Serpil Opperman, 283–90. Bloomington: Indiana University Press, 2014.

———. *Topographies of the Sacred: The Poetics of Place in European Romanticism*. Charlottesville: University of Virginia Press, 2004.

Righter, Robert W. *The Battle over Hetch Hetchy: America's Most Controversial Dam and the Birth of Modern Environmentalism*. New York: Oxford University Press, 2005.

Robsahm, Carl. *Memoirs of Swedenborg and Other Documents*. Edited by Stephen McNeilly. Translated by Anders Hallengren. London: Swedenborg Society, 2011.

Rorty, Richard. *Philosophy and the Mirror of Nature*. Princeton, N.J.: Princeton University Press, 1979.

Ronnberg, Ami, and Kathleen Martin, eds. *The Book of Symbols: Reflections on Archetypal Images*. New York: Taschen, 2010.

Rose, Deborah Bird. "Shimmer: When All You Love is Being Trashed." In *Arts of Living on a Damaged Planet*, edited by Anna Tsing, Heather Swanson, Elaine Gan, and Nils Bubandt, G51–63. Minneapolis: University of Minneapolis Press, 2017.

Rose, Jonathan. "The Ornaments in Swedenborg's Theological First Editions." *Covenant: A Journal Devoted to the Study of the Five Churches* 1 (1998): 293–363.

Rosenberg, Raphael. "Was There a First Abstract Painter? Af Klint's Amimetic Images and Kandinsky's Abstract Art." In *Hilma af Klint: The Art of Seeing the Invisible*, edited by Kurt Almqvist, 87–100. Stockholm: Axel och Margaret Ax:son Johnson's Stiftelse, 2015.

Rowlandson, William. *Borges, Swedenborg, and Mysticism*. Oxford: Peter Lang, 2013.

Rubenstein, Mary-Jane. *Worlds without End: The Many Lives of the Multiverse*. New York: Columbia University Press, 2014.

Sachs, Aaron. *The Humboldt Current: Nineteenth-Century Exploration and the Roots of American Environmentalism*. New York: Penguin, 2007.

Sandilands, Catriona. "The Importance of Reading Queerly: Jewett's *Deephaven* as Feminist Ecology." *ISLE: Interdisciplinary Studies in Literature and Environment* 11, no. 2 (July 1, 2004): 57–77.

Schaffer, Kristen. "Burnham's Plan of Chicago and Universal Human." Paper presented to Swedenborg and the Arts Conference, June 7, 2017, Bryn Athyn College, Pa.

———. *Daniel H. Burnham: Visionary Architect and Planner*. Edited by Scott J. Tilden, with photographs by Paul Rocheleau. New York: Rizzoli, 2003.

Schmidt, Leigh Eric. *Hearing Things: Religion, Illusion, and the American Enlightenment*. Cambridge, Mass.: Harvard University Press, 2002.

Schneck, Peter. "'The Great Secular Transcendence': Don DeLillo and the Desire for Numinous Experience." In *Terrorism, Media, and the Ethics of Fiction: Transatlantic Perspectives on Don DeLillo*, edited by Peter Schneck and Philipp Schweighauser, 204–19. London: Continuum, 2010.

Schuchard, Marsha Keith. *Emanuel Swedenborg, Secret Agent on Earth and in Heaven: Jacobites, Jews, and Freemasons in Early Modern Sweden*. Leiden: Brill, 2012.

———. "Why Mrs. Blake Cried: Swedenborg, Blake, and the Sexual Basis of Spiritual Vision." *Esoterica* 2 (2000): 45–93.

Schuyler, David. *Sanctified Landscape: Writers, Artists, and the Hudson River Valley, 1820–1909*. Ithaca, N.Y.: Cornell University Press, 2012.

Seidman, Naomi. *The Marriage Plot; or, How Jews Fell in Love with Love, and with Literature*. Stanford, Calif.: Stanford University Press, 2016.

Shaw, Sylvia D. "Sampson Reed: Swedenborgian Pioneer in American Literature." MA thesis, Clark University, 1986.

Shelley, Mary Wollstonecraft. *Frankenstein; or, The Modern Prometheus*. Edited by D. L. McDonald and Kathleen Sherf. Ontario, Ont.: Broadview Editions, 2012.

Sigstedt, Cyriel Sigrid. *The Swedenborg Epic: The Life and Works of Emanuel Swedenborg*. New York: Bookman Associates, 1952.

Silver, Ednah. "Sketches of the New Church." Typescript manuscript in Joseph Worcester Files, 72.9. Archives of the Center for Swedenborgian Studies at the Graduate Theological Union, Berkeley, Calif.

Silver, Richard Kenneth. "The Spiritual Kingdom in America: the Influence of Emanuel Swedenborg on American Society and Culture, 1815–1860." PhD diss., Stanford University, 1983.

Sinclair, Iain. *Blake's London: The Topographic Sublime*. London: Swedenborg Society, 2011.

———. *Swimming to Heaven: The Lost Rivers of London*. London: Swedenborg Society, 2013.

Skinner, Jonathan. "What Is Eco-Poetics?" *Jacket2*. Accessed December 27, 2018. http://jacket2.org/commentary/jonathan-skinner.

Skrbina, David. *Panpsychism and the West*. Cambridge, Mass.: MIT Press, 2005.

———. "Panpsychism in History: An Overview." In *Mind That Abides: Panpsychism in the New Millennium*, edited by David Skrbina, 1–29. Philadelphia: John Benjamins Publishing Company, 2009.

"Slaughter in Florida." *Audubon Magazine* 1, no. 6 (July 1887): 139.

Solnit, Rebecca. "Every Corner Is Alive: Eliot Porter as Environmental Artist." In *A Keener Perception: Ecocritical Studies in American Art History*, edited by Alan Braddock and Christoph Irmscher, 225–53. Tuscaloosa: University of Alabama Press, 2009.

———. *Savage Dreams: A Journey into the Hidden Wars of the American West*. Berkeley: University of California Press, 2014.

Spence, Mark David. *Dispossessing the Wilderness: Indian Removal and the Making of the National Parks*. New York: Oxford University Press, 2000.

Stengel, Friedemann. *Aufklärung bis zum Himmel: Emanuel Swedenborg im Kontext der Theologie und Philosophie des 18. Jahrhunderts* [Enlightenment all the way up to heaven: Emanuel Swedenborg in the theological and philosophical contexts of the 18th century]. Tübingen: Mohr Siebeck, 2011.

———. "Rezension über Marsha Keith Schuchard, 'Emanuel Swedenborg, Secret Agent on Earth and in Heaven.'" *Journal of Nordic Studies* 7, no. 1 (2013): 86–129.

Stengers, Isabelle. *Thinking with Whitehead: A Free and Wild Creation of Concepts*. Translated by Michael Chase. Cambridge, Mass.: Harvard University Press, 2011.

Stoll, Mark. *Inherit the Holy Mountain: Religion and the Rise of American Environmentalism*. New York: Oxford University Press, 2015.

———. "Milton in Yosemite: *Paradise Lost* and the National Parks Idea." *Environmental History* 13, no. 2 (2008): 237–74.

Stone, Christopher D. *Should Trees Have Standing?: Law, Morality, and the Environment*. New York: Oxford University Press, 2010.

Strong, Lisa. "Mount Corcoran." In *Corcoran Gallery of Art: American Paintings to 1945*, edited by Sarah Cash, 140–41. Washington, D.C.: Corcoran Gallery of Art, 2012.

Suzuki, Daisetz Taitaro. *Swedenborg: Buddha of the North*. Translated by Andrew Bernstein. West Chester, Pa.: Swedenborg Foundation, 1996.

Swedenborg, Emanuel. *The Animal Kingdom, Considered Anatomically, Physically, and Philosophically*. Translated by James John Garth Wilkinson. 2 vols. London: Newbery, 1843.

———. *The Apocalypse Revealed*. Translated by John Whitehead. New York: Swedenborg Foundation, 1949.

———. *Arcana Coelestia*. Edited by John Faulkner Potts. Translated by John Clowes. Rev. ed. 12 vols. West Chester, Pa.: Swedenborg Foundation, 1998.

———. *Catalogus Bibliothecae Emanuelis Swedenborgii*. Edited by Alfred Stroh. Stockholm: Aftonbladets Tryckerei, 1907.

———. *Divine Love and Wisdom*. New Century Edition. Translated by George F. Dole. West Chester, Pa.: Swedenborg Foundation, 2010.

———. *Divine Providence*. New Century Edition. Translated by George F. Dole. West Chester, Pa.: Swedenborg Foundation, 2010.

———. *The Earthlike Bodies Called Planets in Our Solar System and in Deep Space, Their Inhabitants, and the Spirits and Angels There: Drawn from Things Heard and Seen*. Translated by George F. Dole. In *The Shorter Works of 1758*, 423–523. New Century Edition. West Chester, Pa.: Swedenborg Foundation, 2018.

———. *Emanuel Swedenborg's Diary: Recounting Spiritual Experiences during the Years 1745 to 1765*. Translated by Durban Odhner. Bryn Athyn, Pa.: General Church of the New Jerusalem, 1998.

———. *Heaven and Its Wonders and Hell: Drawn from Things Seen and Heard*. Translated by George F. Dole. West Chester, Pa.: Swedenborg Foundation, 2000.

———. *The Interaction between Soul and Body: Which Is Believed to Be Either by Physical Influx, or Spiritual Influx, or by Preestablished Harmony*. In *Miscellaneous Theological Works*, 318–51. Translated and edited by John Whitehead. West Chester, Pa.: Swedenborg Foundation, 1996.

———. *The Letters and Memorials of Emanuel Swedenborg*. 2 vols. Edited by Alfred Acton. Bryn Athyn, Pa.: Swedenborg Scientific Association, 1955.

———. *Ludus Heliconius: And Other Latin Poems*. Translated by Hans Helander. Series Acta Universitatis Upsaliensis no. 23. Uppsala: Uppsala University, 1995.

———. *A Philosopher's Notebook*. Translated by Alfred Acton. Bryn Athyn, Pa.: Swedenborg Scientific Association, 2009.

———. *The Philosophy of the Infinite; or, Outlines of a Philosophical Argument on the Infinite*. Translated by James John Garth Wilkinson. Boston: Otis Clapp, 1848.

———. *The Principia; or, The First Principles of Natural Things, Being New Attempts toward a Philosophical Explanation of the Elementary World*. Translated by Augustus Clissold. 2 vols. London: W. Newbery, 1845.

———. *The Schmidius Marginalia Together with the Expository Material of the Index Biblicus*. Translated and edited by E. E. Iungerich. Bryn Athyn, Pa.: Academy of the New Church, 1917.

———. *Some Specimens of a Work on the Principles of Chemistry, with Other Treatises*. Translated by Charles Edward Strutt. London, W. Newbery, 1847.

———. *Swedenborg's Dream Diary*. Translated by Lars Bergquist. West Chester, Pa.: Swedenborg Foundation, 2001.

———. *A Theosophic Treatise on the Nature of Influx: As It Respects the Communication and Operations of Soul and Body*. Translated by Robert Hindmarsh. London: R. Hindmarsh, 1784.

———. *True Christianity*. The New Century Edition. Translated by Jonathan Rose. 2 vols. West Chester, Pa.: Swedenborg Foundation, 2010.
———. *The Worship and Love of God: In Three Parts*. Translated by Alfred H. Stroh and Frank Sewall. West Chester, Pa.: Swedenborg Foundation, 1996.
Swift, Ann. *Bubbling Springs and Bushy Dells: The Story of Piedmont Springs Park*. Piedmont, Calif.: Piedmont Historical Society, 1998.
Tafel, Rudolph. *Bischof Jesper Swedberg: Der Vater Emanuel Swedenborgs: Eine Biographie* [Bishop Jesper Swedberg: The father of Emanuel Swedenborg: A biography]. Frankfurt am Main: J. G. Mittnacht, 1880.
———. *Documents Concerning the Life and Character of Emanuel Swedenborg*. 3 vols. London: Swedenborg Society, 1875.
Taylor, Andrew. *Henry James and the Father Question*. Cambridge: Cambridge University Press, 2009.
Taylor, Bron. *Dark Green Religion: Nature Spirituality and the Planetary Future*. Berkeley: University of California Press, 2010.
———. "The Greening of the Religion Hypothesis (Part One)." *Journal for the Study of Religion, Nature, and Culture*, 10, no. 3 (2016): 268–305.
———. "The Religion and Politics of Earth First!" *The Ecologist* 21, no. 6 (1991): 258–66.
Taylor, Charles. *Modern Social Imaginaries*. Durham, N.C.: Duke University Press, 2004.
———. *A Secular Age*. Stanford, Calif.: Stanford University Press, 2007.
Taylor, Eugene. "Peirce and Swedenborg." *Studia Swedenborgiana* 6, no. 1 (1986): 25–51.
Taylor, Mark C. *Refiguring the Spiritual: Beuys, Barney, Turrell, Goldsworthy*. New York: Columbia University Press, 2012.
Thomas, John L. *Alternative America: Henry George, Edward Bellamy, Henry Demarest Lloyd, and the Adversary Tradition*. Cambridge, Mass.: Belknap Press of Harvard University Press, 1983.
Thoreau, Henry David. *Journal*, Vol. 4: *1851–1852*. Edited by Leonard N. Neufeldt and Nancy Craig Simmons. Princeton, N.J.: Princeton University Press, 1992.
Tobish, Othmar. *The Garden Church of San Francisco, a Church of the New Jerusalem, Swedenborgian: A Description of Its Symbolisms, Art Treasures, and Excerpts from Historical Documents*. Pamphlet, 1964. Archives of the San Francisco Swedenborgian Church.
Traubel, Horace. *With Walt Whitman in Camden*, Vol. 5: *April 8–September 14*, 1889. Edited by Gertrude Traubel. Carbondale: Southern Illinois University Press, 1964.
Tuan, Yi-fu. *Topophilia: A Study of Environmental Perception, Attitudes, and Values*. Englewood Cliffs, N.J.: Prentice-Hall, 1974.
Turner, Victor. *Dramas, Fields, and Metaphors: Symbolic Action in Human Society*. Ithaca, N.Y.: Cornell University Press, 1975.

Tweed, Thomas A. “American Occultism and Japanese Buddhism: Albert J. Edmunds, D. T. Suzuki, and Translocative History.” *Japanese Journal of Religious Studies* 32, no. 2 (2005): 249–81.

Uexküll, Jacob von. “A Stroll through the Worlds of Animals and Men: A Picturebook of Invisible Worlds, 1934.” In *Life Itself: On the Question of What It Essentially Is; Its Materialities, Its Characteristics, Considering That the Attempts to Answer This Question by Occidental Sciences and Philosophies Have Proven Unsatisfactory*, edited by Daniel Birnbaum, Stefanie Hessler, Carsten Höller, and Jo Widoff, 32–38. Cologne: Walther König Verlag, 2016.

Uglow, Jenny. *The Lunar Men: The Friends Who Made the Future, 1730–1810*. London: Faber and Faber, 2002.

Versluis, Arthur. *The Esoteric Origins of the American Renaissance*. Oxford: Oxford University Press, 2001.

Viscomi, Joseph. “In the Caves of Heaven and Hell: Swedenborg and Printmaking in Blake's *Marriage*.” In *Blake in the Nineties*, edited by David Worrall and Steven H. Clark, 27–60. New York: St. Martin's Press, 1999.

Voelz, Johannes. *Transcendental Resistance: The New Americanists and Emerson's Challenge*. Hanover, N.H.: Dartmouth University Press, 2010.

Vogel, Carol. “New Holdings in the Modern,” *New York Times*, November 12, 2004.

Walker, Colette. “The Language of Form and Color: Traces of Swedenborg's Doctrine of Correspondences in Kandinsky's Concerning the Spiritual in Art.” *The New Philosophy: Journal of the Swedenborg Scientific Association* 117, no. 3/4 (December 2014): 165–86.

Wallace-Wells, David. *The Uninhabitable Earth: Life after Warming*. New York: Tim Duggan Books, 2019.

Wallach, Alan. “Some Further Thoughts on the Panoramic Mode in Hudson River School Landscape Painting.” In *Within the Landscape: Essays on Nineteenth-Century American Art and Culture*, edited by Philip Earenfight and Nancy Siegel, 99–128. University Park: Pennsylvania State University Press, 2005.

———. “Wadsworth's Tower: An Episode in the History of American Landscape Vision.” *American Art* 10, no. 3 (October 1, 1996): 9–27.

Walls, Laura Dassow. *Emerson's Life in Science: The Culture of Truth*. Ithaca, N.Y.: Cornell University Press, 2003.

———. *Henry David Thoreau: A Life*. Chicago: University of Chicago Press, 2017.

———. *The Passage to the Cosmos: Alexander von Humboldt and the Shaping of America*. Chicago: University of Chicago Press, 2011.

———. *Seeing New Worlds: Henry David Thoreau and Nineteenth-Century Natural Science*. Madison: University of Wisconsin Press, 1995.

Weber, Max. *From Max Weber: Essays in Sociology*. Translated and edited by H. H. Girth and C. Wright Mills. New York: Oxford University Press, 1981.

Weintraub, Linda. *To Life!: Eco Art in Pursuit of a Sustainable Planet*. Berkeley: University of California Press, 2012.

Westfahl, Arline. *One Person Can Make a Difference: A Story of Paul Kroegel and Pelican Island*. Vero Beach, Fla.: Indian River County Historical Society, 2003.

Wheeler, Wendy. "Natural Play, Natural Metaphor, and Natural Stories: Biosemiotic Realism." In *Material Ecocriticism*, edited by Serenella Iovino and Serpil Oppermann, 67–79. Bloomington: Indiana University Press, 2014.

———. *The Whole Creature: Complexity, Biosemiotics, and the Evolution of Culture*. London: Lawrence and Wishart, 2006.

White, Lynn. "The Historical Roots of Our Ecological Crisis." *Science* 155, no. 3767 (March 1967): 1203–7.

Whitehead, Alfred North. *Process and Reality: An Essay in Cosmology*. New York: Free Press, 1978.

———. *Science and the Modern World*. New York: the Free Press, 1967.

Whitman, Walt. *Poetry and Prose*. New York: Library of America, 1996.

———. *The Portable Walt Whitman*. Edited by Michael Warner. New York: Penguin Classics, 2003.

"Wholesale Destruction of Birds in Florida," *Audubon Magazine* 1, no. 8 (September 1887): 178–80.

Wilkinson, James John Garth. "Correspondence." In *Aesthetic Papers*, edited by Elizabeth Peabody, 112–45. New York: G. P. Putnam, 1849.

———. Correspondence from James John Garth Wilkinson to Henry James Sr., Ralph Waldo Emerson, and others. K–123, E 1–8, K–124a. Swedenborg Society, London.

———. *Emanuel Swedenborg: A Biography*. London: W. Newbury, 1849.

———. "Introductory Remarks." In *The Philosophy of the Infinite; or, Outlines of a Philosophical Argument on the Infinite*, translated by James John Garth Wilkinson, 3–9. Boston: Otis Clapp, 1848.

Wilkinson, Lynn. *The Dream of an Absolute Language: Emanuel Swedenborg and French Literary Culture*. Albany: State University of New York Press, 1996.

Williams, Dennis. "John Muir, Christian Mysticism, and the Spiritual Value of Nature." In *John Muir: Life and Work*, edited by Sally M. Miller, 64–99. Albuquerque: University of New Mexico Press, 1990.

Wilson, Eric. *The Spiritual History of Ice: Romanticism, Science, and the Imagination*. New York: Palgrave Macmillan, 2003.

Wilson, William. *The City Beautiful Movement*. Baltimore: Johns Hopkins University Press, 1994.

Wittgenstein, Ludwig. *Philosophical Investigations*. 4th ed. Translated by G.E.M. Anscombe. Malden, Mass.: Wiley-Blackwell, 2009.

Wohlleben, Peter. *Das geheime Leben der Bäume: Was sie fühlen, wie sie kommunizieren—die Entdeckung einer verborgenen Welt* (The hidden life of trees: What they feel, how they communicate). Munich: Ludwig Buchverlag, 2015.

———. *The Hidden Life of Trees: What They Feel, What They Communicate—Discoveries from a Secret World*. Translated by Jane Billinghurst. Vancouver: Greystone Books, 2016.

Wolfe, Cary. *What Is Posthumanism?* Minneapolis: University of Minnesota Press, 2009.

Wolfe, Linnie Marsh. *Son of the Wilderness: The Life of John Muir.* Madison: University of Wisconsin Press, 2003.

Worcester, Alfred. "The Rev. Joseph Worcester: A Memoir and Extracts from His Letters." Unpublished typescript, n.d. Archives of the San Francisco Swedenborgian Church.

Worster, Donald. *A Passion for Nature: The Life of John Muir.* Oxford: Oxford University Press, 2011.

Wurtz, Michael. "John Muir Disrupts a Séance." *John Muir Newsletter* 17, no. 1 (Winter 2006/2007): 2–4.

Wulf, Andrea. *The Invention of Nature: Alexander von Humboldt's New World.* New York: Knopf, 2015.

Zuber, Devin. "An Apocalypse of Mind: Cracking the Jerusalem Code in Emanuel Swedenborg's Theosophy." In *Tracing the Jerusalem Code: Christian Cultures in Scandinavia,* edited by Anna Bohlin and Ragnhild Zorgati. Turnhout, Belgium: Brepols Publishing, 2019.

———. "Buddha of the North: Swedenborg and Transpacific Zen." *Religion and the Arts* 14, no. 1/2 (2010): 1–33.

———. "Can Beauty (Still) Save the Planet?: Eco-Aesthetics for the Anthropocene." In *Gestures to the Divine: the Art of Hagit Cohen,* edited by Elizabeth Pena, 16–21. Berkeley, Calif.: Center for the Arts and Religion, 2018.

———. "Flânerie at Ground Zero: Aesthetic Countermemories in Lower Manhattan." *American Quarterly* 58, no. 2 (2006): 269–99.

———. "The Reservation Paints Back: Native American Aesthetics, Law, and Imagetext." In *Law and Culture: Methods, Concepts, Approaches,* edited by Peter Scheck and Sabine Meyer. Amsterdam: De Gruyter, 2019.

———. "Satanic Mills and Jerusalem Redeemed: William Blake's Urban Ecopoetics." *Green Letters* 10 (Winter 2008–9): 39–49.

———. "Spiritualized Science and the Celestial Artist: Nathaniel Hawthorne and Swedenborgian Aesthetics." In *Philosophy, Literature, Mysticism,* edited by Stephen McNeilly, 175–216. London: Swedenborg Society, 2013.

———. "'Those Almost Uninhabited Shores': John James Audubon and the Violence of Space." In *The Audubon Trilogy: Delineations of American Scenes and Manners,* DVD with chapbook essay. Manchester: Cornerhouse Gallery Publications, 2010.

Index

Recent books in the series

Studies in Religion and Culture

A Language of Things: Emanuel Swedenborg and the American Environmental Imagination
Devin P. Zuber

The Pragmatist Turn: Religion, the Enlightenment, and the Formation of American Literature
Giles Gunn

Rethinking Sincerity and Authenticity: The Ethics of Theatricality in Kant, Kierkegaard, and Levinas
Howard Pickett

The Newark Earthworks: Enduring Monuments, Contested Meanings
Lindsay Jones and Richard D. Shiels, editors

Ideas to Live For: Toward a Global Ethics
Giles Gunn

The Pagan Writes Back: When World Religion Meets World Literature
Zhange Ni

Freud and Augustine in Dialogue: Psychoanalysis, Mysticism, and the Culture of Modern Spirituality
William B. Parsons

Vigilant Faith: Passionate Agnosticism in a Secular World
Daniel Boscaljon

Postmodernism and the Revolution in Religious Theory: Toward a Semiotics of the Event
Carl Raschke

Textual Intimacy: Autobiography and Religious Identities
Wesley A. Kort

When the Sun Danced: Myth, Miracles, and Modernity in Early Twentieth-Century Portugal
Jeffrey S. Bennett

Encountering the Secular: Philosophical Endeavors in Religion and Culture
J. Heath Atchley